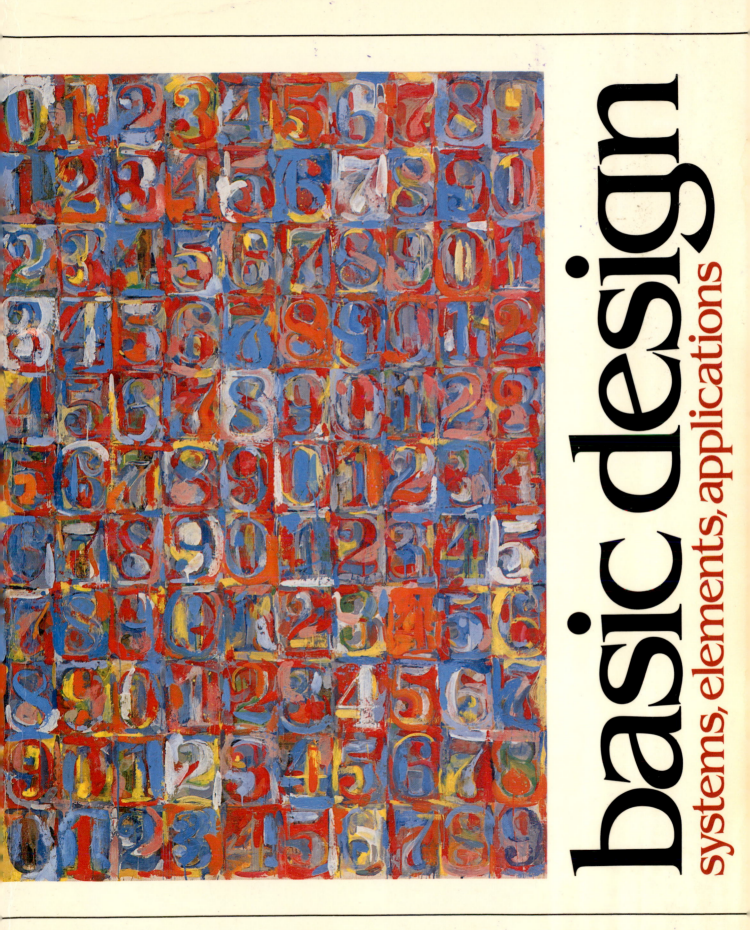

basic design

systems, elements, applications

basic design

systems, elements, applications

The authors (left to right): Michael J. Smith, John Adkins Richardson, Floyd W. Coleman

basic design
systems, elements, applications

John Adkins Richardson
Floyd W. Coleman
Michael J. Smith

Southern Illinois University at Edwardsville

Prentice-Hall, Inc., Englewood Cliffs, New Jersey 07632

Library of Congress Cataloging in Publication Data

RICHARDSON, JOHN ADKINS.
  Basic design.

  Bibliography: p.
  Includes index.
  1. Design.   I. Coleman, Floyd W.
II. Smith, Michael J. III. Title.
NK1510.R44   1984   745.4   82-13289
ISBN 0-13-060186-1

Editorial/production supervision: Fred Bernardi
Interior and cover design: Judith A. Matz
Manufacturing buyer: Harry P. Baisley
Page layouts: Caliber Design Planning, Inc.
Cover photo: Jasper Johns. *Numbers in Color.* 1958–59. En-
              caustic and collage on canvas, 5′7″ ×
              4′1/2″. Albright-Knox Art Gallery, Buf-
              falo, N.Y.

Printed in the United States of America

10 9 8 7 6 5 4 3 2 1

ISBN: 0-13-060186-1

PRENTICE-HALL INTERNATIONAL, INC., *London*
PRENTICE-HALL OF AUSTRALIA PTY. LIMITED, *Sydney*
EDITORA PRENTICE-HALL DO BRASIL, LTDA., *Rio de Janiero*
PRENTICE-HALL CANADA INC., *Toronto*
PRENTICE-HALL OF INDIA PRIVATE LIMITED, *New Delhi*
PRENTICE-HALL OF JAPAN, INC., *Tokyo*
PRENTICE-HALL OF SOUTHEAST ASIA PTE. LTD., *Singapore*
WHITEHALL BOOKS LIMITED, *Wellington, New Zealand*

For Glenda, Yvonne, and Valerie

contents

foreword

None of the authors seems to have any idea of who reads prefaces and forewords or what sort of instructor demands that students do. One of us always does read them and another puts questions about them on midterm examinations. Obviously, *you* are reading this one, at least. Welcome. What we want to deal with in this opening, principally, are our intentions.

We undertook to compose this text because we were discontent. No book dealing with the fundamentals of design seemed to be entirely satisfactory for the introductory course we teach, and the textbooks we have tried to use have provided little foundation for the wide range of studies to which beginning design instruction is prerequisite. Fledgling painters, sculptors, ceramists, graphic designers, industrial designers, architects, interior designers, weavers, Conceptualists, printmakers, *et al.* all study basic design. Some of the design textbooks we used were too concerned with theoretical underpinnings and the psychology of perception; others were too unconcerned with thought. Some were fine for the applied arts and others were completely antagonistic to anything apart from noble undertakings from which none but the occasional genius can earn a living. Most disconcerting of all, to the writers, was the tendency of the *better* books to dismiss as unworthy of serious consideration anything that has come down from the past. Only the shallower and less well-informed books dealt with older orders of vision as something other than symbols of some wearisome debacle called ''art history.'' It did not seem to any of us that irrevocable alienation from the past was a necessary condition for design education.

There was really nothing unusual in our disappointment. Dissatisfaction with available texts is common among those who deal with basic design. They are always trying to solve the deficiencies of the market by adding to it books that, upon publication, are found wanting by other design teachers who, in turn, are inspired to write still more books. The one good thing about this situation—at least from an author's vantage point—is that it seems to provide an audience for renewed attempts. And some of those books, of course, do fill the needs of some people in the field. Thus, Richardson/Coleman/Smith enter into the vortex of the great design text-turnover. To those of our readers who may at some time be induced to replace *Basic Design: Systems, Elements, Applications* with something that seems to them more rational, more relevant, rangier, or more rousing, we proffer a well-meant admonition.

It seemed to us, when we commenced to compose this text, that it should be an easy thing. Not only did we have experience as teachers of design at colleges and universities all over the nation, we have, collectively, a remarkably wide range of professional experiences as writers, painters, draftsmen, graphic designers, and printmakers, variously experienced in art history, home design, technical illustration, cartooning, advertising, and even concrete finishing, banjo making, and housekeeping. Best of all, we are by no means in ideological accord on economics, politics, aesthetics, or pedagogy, and yet we have learned to work in harmony, agreeing to disagree. That is, we are friends who argue about nuclear power, welfare priorities, affirmative action, Conceptual Art, consumer rights, advertising, ecological significances, Third World realities, income distribution, merit pay, and so on. It seemed as if the merger of a common objective with diverse attitudes might produce an ideal synthesis, satisfying every need. Alas, it was not to be. For the "perfect" design text, showing every side of all contentions and each school of thought, turned out to be a massive volume the size of three books about as big as the typical survey of art history. Such a definitive tome might be worth having for a reference work, but it would be far too large and costly for a text. What you have in hand, however, is not a scaled-down facsimile of our overweight ideal. It is an alternative to it.

We have tried to accommodate all of the things we think are essential in an introduction to design and tried to include, as well, a few links with various specialties. All italicized terms either are defined upon their first appearance and indexed accordingly, or are in the glossary at the end of the book. Beginning with a chapter on the attitudinal or "philosophical" schisms, which are very pronounced in the field of design, we have tried to outline, as candidly, starkly, and indelicately as possible, some of the genuine problems faced by artists and designers in the world today.

Part II of the book, entitled "Systems," presents a variety of fairly standard approaches to design. Some of these are, however, quite different from what are customarily found in basic design textbooks; they pertain to pictorial composition and are among the techniques that have been used by commercial illustrators for at least a century. For a long while they have been notable for their absence from design study—as, for that matter, has illustration itself. Since even the fine arts today entail much more realistic imagery than in the four decades just past, and since there are sometimes chasm-like disparities between photographic imagery and representational paintings, it is appropriate to consider similarities and differences among photographs, paintings, and reality and between traditional composition and modern design, a consideration which inevitably involves some art historical explication.

In Part III we have taken up the so-called "elements" of art and design. Perhaps detached logic would seem to dictate that the elements be considered first, inasmuch as they are supposed to constitute a "language of design." But, in fact, it is rare to introduce them at the beginning of a book, for the same reason children aren't made to study the parts of speech until after they have been using subjects, verbs, adjectives, and adverbs for a number of years. However that may be, Chapters 5 through 8 show how the elements can be organized and interrelated in different kinds of designs for various purposes through sundry media.

Finally, in the last part of the book we take up some applications of design principles in connection with graphic design, domestic architecture, and Conceptual Art. Of all the possible areas we might have chosen to focus attention upon, why did we chose those three? Why not industrial design, urban planning, and pure performance art? Why not serious painting or printmaking? Why not textiles and packaging and computer graphics? Well, naturally, our selections were dictated in part by overlapping personal interests and priorities, but there are other reasons besides. We could certainly have picked other topics, in which some of us are more expert than we are collectively in the ones we decided to use. Graphic design—generally, what used to be called "commercial art"—not only has special, immediate vocational potential, but is something that everyone in art or art education needs, sooner or later, to know a least a little bit about. Art exhibitions and similar events require posters, brochures, flyers, and labels. Even if one is not required to produce these, it is more often than not helpful to know something about the structure of lettering styles and the possible uses of major printing processes. As for domestic architecture, dwelling space is, as we write, at an increasingly chronic premium—a condition that is likely to become more and more critical as time wears on. A related issue involves designing houses that conserve energy, and, as it happens, some of the most advanced work in that area is being done locally, by a man and woman well-known to the authors. Last, Conceptual Art can be said to be the *ultimate* design alternative to the rather down-to-earth approach stressed throughout the rest of the book. In a sense, two of us have compromised a general disdain for what we suspect is often undertaken by charlatans and fools out of respect for the third author, who feels that this book tends to pander to the status quo so far as its definition of design is concerned.

Each chapter of the book, beginning with Chapter 2, contains some exercises that we have used in teaching design and have found to be effective. Wherever possible and pertinent, we have illustrated these with things created by students. Of course, we did not strive to represent the entire range of quality, all the way from excellent to complete failure, but neither is what you see reproduced herein the best of the best. Why not? Because we couldn't always get our hands on what was best. Often, the superb work had vanished along with the students, and attempts to reach them came to naught. Just understand that the examples are only that—examples—and are intended neither to represent typical products nor establish a mark of excellence. Too, there is a peculiar kind of dynamic in every studio classroom; an exercise might work for one instructor in a given class and not work for other instructors or for the same instructor with different groups. All we can say is that we have selected what worked consistently for at least two of the authors. We have also included those we just could not bring ourselves to leave out, even when it overburdened a chapter.

We do welcome feedback and suggestions. Color reproductions, of which people always want more, are expensive and add to the unit cost of a book. Furthermore, since it takes about two years to bring out a book, once it has been finished in manuscript, it can never be as up-to-date as most of us would wish. Some things just cannot be the way we would like. On the other hand, some could be the way you might prefer, were we given the advantage of your experience, imagination, enthusiasm, or annoyance. Corrections are also taken in with gratitude. It isn't possible to write a book like this without making at least a few errors. Of course, we'd like to blame all of them on inept editors and copy readers, but it is considered bad form to suggest that mistakes can be those of anyone except the authors. And, for a fact, whatever is banal, uninformed, or preposterous belongs to one or another of the authors and anything that seems original, pretty good, and full of insight is doubtless due to one of the other two, who themselves must give credit to a fair-sized crowd of people.

At Prentice-Hall, Executive Editor of Humanities, Norwell Therien, Jr., and his assistant, Jean Wachter, pressed us from pent-up lethargy to modestly effective action and also inspired some essential modifications of the book. Fred Bernardi, in charge of the actual production, has produced a fine volume despite having had to overcome the many confusions and pesky delays that arose from things over which none of us had control. The design itself is the work of Judith A. Matz. Copy editor Ilene McGrath brought consistency to a text reflecting three different points-of-view and moderated some of the excesses from which no book on design seems ever to be entirely free.

Without the assistance of artists, architects, designers, museums, galleries, publishers, agencies, manufacturers, and photographic services, no book of this kind could be produced. Although they are recognized elsewhere in the captions and photographic credits, we offer here our special thanks.

Locally, our colleagues have been highly supportive in many ways, and our students more tolerant than could have been imagined. At Southern Illinois University-Edwardsville, Vaughnie Lindsay, Dean of the Graduate School, Hollis White, Dean of Fine Arts and Communications, and Don Davis, Chairperson of the Department of Art and Design, were notable, among administrators, for their moral and occasionally material support. As always, Charles Cox and SIUE Photographic Services were first-rate. We had also, however, the help of photographer/designer Robert Weaver without whose skills and gratuitous expenditure of time and energy the visual range of examples would be seriously wanting. Prof. David C. Huntley, Director of Cultural Arts and University Museums, was notably responsive to requests for special photographs of works in the SIUE collection. Fine Arts Librarian, Philip Calcagno, provided critical assistance at several points. Finally, we must give our collective recognition to the secretary of Art and Design, Rosemary Drew, whose efficiency, good humor, and forebearance made the preparation of this book a great deal less onerous than it might otherwise have been.

part 1

beginnings

chapter 1

design, designers, and art

Books on what is called *design* nearly always open with an apology. The subject is so comprehensive, concerned as it is with the structure and organization of everything from microscopic nature (Fig. 1) to city planning (Fig. 2) and from typography (Fig. 3) to trusswork (Fig. 4), that it is impossible to cover in a hundred volumes, let alone in one. Often, writers attempt to show that what is called ''art'' is just another of the many instances in which design principles have found expression. Fair enough, but if you read much about the subject, you will soon realize that such writing reveals a kind of guarded defensiveness, almost as though the exponents of design theory expected the painters and sculptors they cite to rise up and disclaim the connections. There is, indeed, a certain tension between those who practice what we commonly refer to as ''fine art'' and those whose works are examples of what used to be called ''applied art.'' The whole thing has to do with social status and we might as well deal with it straight away and be done with it.

In human society it is generally the rule that what is rare, precious, and unique commands a higher value than what is commonplace and that the creator of the one-of-a-kind luxury item (such as an oil painting) is going to enjoy greater social status than an equally successful craftsperson whose products are multiplied by machinery for the masses. Of course, famous commercial designers are more celebrated than are unknown fine artists, and our ability today to reproduce a Rembrandt or Picasso in full color raises some complicated issues. However, the fact remains that one-of-a-kind objects carry a status that even the most exquisite and useful commonplace objects will not. In the same way, a limited-edition, hand-printed etching is of greater value than a poster run off in thousands of copies. The upshot is that today art work done for itself, without a specific purpose apart from an artist's need for self-expression, is highly esteemed. Art work of equal quality done for a practical purpose or for reproduction may be greatly admired but it is not going to be worshipped in the way the unique art object is. Weber's cartoon (Fig. 5) says it all.

In one sense, all art is applied art. That is, the most spiritually motivated painter creates objects that can be used to ornament walls. The intentions of the artist have nothing to do with this. That Vincent van Gogh (1853–1890) painted pictures to fulfill a deep, compelling hunger for personal expression is the most obvious fact about his life. He sold practically nothing during his lifetime; his success and fame were posthumous; yet today his works hang in museums and decorate mansions. Even he used the paintings as part of the decor of his unpretentious little house in Arles, France. So, in a way, the van Gogh over van Gogh's bed (see Fig. 6) was as much a part of the applied arts as a drapery or rug would have been. Of course, we are not equating the activities of painters with those of textile designers just because a painting can serve as ornament as well as a means of self-expression. After all, the designer of a drapery pattern may receive much more gratification from the self-expression involved in its creation than from doing an oil painting even though the presumed purpose of the fabric is to decorate a home rather than to express a viewpoint.

Nonetheless, the fact that high art *can* be used to decorate homes, and can therefore be treated just like any other commodity by the wealthy and status-conscious, has led certain contemporary artists, who think this too materialistic and crass, to create art

FIGURE 1 Leaf

FIGURE 2 Airview. Southern Illinois University at Edwardsville. Photo: John Rendleman

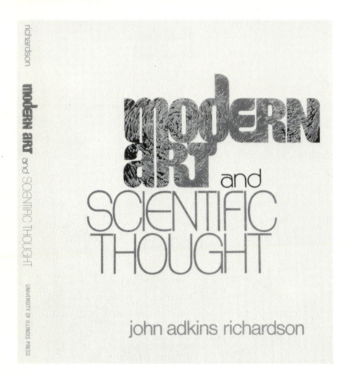

FIGURE 3 CHARLES HUGHES. Book jacket for *Modern Art and Scientific Thought* by John Adkins Richardson. © 1971 by the Board of Trustees of the University of Illinois.

FIGURE 4 Trusswork

FIGURE 5 ROBERT WEBER. © 1975, The New Yorker Magazine, Inc.

"Well, I like Titian, and Rubens, and, of course, El Greco. But Delacroix is my god."

FIGURE 6 VINCENT VAN GOGH. *Bedroom at Arles.* 1888. Oil on canvas, 28½ × 36". Art Institute of Chicago, Helen Birch Bartlett Memorial Collection

Although van Gogh's oil paintings were done from an intense need to reveal his sense for reality as a merger of the objective facts of the material world with one's subjective reaction to them, he also felt that the pictures were things of beauty which could be used to embellish the world. Here, in his bedroom at Arles, he shows art works hanging on the walls.

works that consist of ideas instead of objects. However, even their efforts, called *Conceptual Art,* cannot escape the commercialism of our age. Since their undertakings are necessarily odd ones and, since the people themselves tend to be both sincere and rather entertaining, Conceptual Art has become *chic* and things connected with it are purchased by collectors in much the same way some people buy stamps, autographs, and old comic books.

There is one obvious distinction between fine artists and applied artists, however. It resides in our expectations of them. We may suspect a given fine artist of being a charlatan in fact. We may, for instance, be skeptical of a woman who says she is dedicated to nothing less than the generation of a new, purely feminine form of artistic expression; we may think that her claim is mere posturing and that she has her idealist's eye fixed firmly on headlines and her realist's eye on the dollars fame most often brings. We are never suspicious in this way of furniture designers, advertising artists, or fashion stylists. Why not? Because we expect such people to work primarily for gain, just like anyone else. And, like anyone else—like tool and die makers, seamstresses, electricians, physicians, gardeners—they may also be completely dedicated to the perfecting of their work in the same way those artists starving in garrets are supposed to be.

So much for distinctions between artists and designers or arts fine and applied. In the main, despite enormous differences, what applies in one area can, with sometimes a little stretching, be carried over into the other. At a more advanced level one might wish to bear in mind what renowned designer Milton Glaser has said: "In design there is a given body of information to be conveyed. . . . That objective is primary in most design activities. On the other hand, the essential function of art is to change and intensify one's perception of reality."[1]

the nature of the book

This book does not hope to achieve some startling "breakthrough" in helping people to conceptualize things in unprecedented ways. People cannot be taught creativity in any meaningful sense. All that can be done through education is to provide some techniques for opening up new possibilities. We shall set forth projects that (1) require the exercise of the imagination and (2) can be relied upon as "backups" and even "crutches." The authors have experimented, over the years, with many different approaches to design and art appreciation and some of the most effective are included here. Some are original with us (at least so far as we know) and some

FIGURE 7 N.C. WYETH. "I said goodbye to Mother..." Illustration for *Treasure Island* by Robert Louis Stevenson (New York: Charles Scribner's Sons, 1911, reissued, 1982). Reprinted with permission of Charles Scribner's Sons © 1911, 1939.

are old standbys. Some are old-fashioned; in fact some of them have not appeared in this kind of book for the past forty years. They are part of the compositional gimmickry of traditional illustration (Fig. 7) and traditional illustration is practically extinct today, pretty much relegated to paperback book covers, comic strips, and odd bits of advertising. However, to ignore it would be to ignore an area of graphic art that has close ties to the past and that appeals to more people more directly than anything outside of television and the movies.

nordic modern and post-modernist attitudes toward design

For a great many years the field of design education was completely dominated in the United States and Europe by a galaxy of ideas, theories, notions, practices, and examples that we will term *Nordic Modern* because its most important sponsors were a German school of industrial design and Swiss typographers who concentrated on clarity and precision in communications.

The school, called *Der Bauhaus* (the building-house), was established in Weimar in 1919 by Walter Gropius and, having proved too radical for that city, moved in 1925 to Dessau, where it continued until 1933, when it was closed by the Nazi government. Many of its staff emigrated to the United States, where they became important spokesmen for the Bauhaus philosophy. At the Institute of Design in Chicago (now the Illinois Institute of Technology), especially, Bauhaus principles were promulgated by Laszlo Moholy-Nagy and his disciples. Universally, the Bauhaus has been recognized as the most influential institution of its type.

At the root of anything the Bauhaus did was "functionalism," the theory that an object will automatically become attractive if it is designed to fulfill its purposes with maximum utility, ignoring all precedents and habits of construction. In traditional schools of architecture and design during the 1920s and 1930s students learned to imitate and modify existing forms from the distant and more recent past. However, students at the Bauhaus and its later offspring were given bizarre problems that challenged their imaginations and forced them to undertake experiments with basic materials. They made springs of wood, structured a single sheet of typing paper so it would support a 25-pound weight one inch above the floor, proposed better-than-igloo domiciles for Eskimos, created paintings designed to "modulate" light rather than merely reflect it, and so on. The graduates of the institution were craftspeople as well as artists. Their training made them more confident of their inventiveness than most designers were.

A great many fine pieces of furniture were produced by the Bauhaus philosophy (see Fig. 8), and what is perhaps the most comfortable and effective chair design of the last century, Charles Eames's "Mister Chair" (Fig. 9), is a product of the same kind of thinking. For what we think of as Bauhaus is not a style or even a curriculum so much as it is an attitude. That is, it was not the purpose of the men and women who taught in Weimar, Dessau, Chicago, and elsewhere to produce work that had a certain kind of "look." They merely wanted to use modern industrial materials and techniques to create things that would work well for human purposes in human society. As it turned out, however, the austere and rather scientific logic of the Bauhaus approach tended to result in buildings, furniture, pictures, and posters that were spare, pristine, neat, and rather anonymous. That had been true from the very outset. The founders of the school had been markedly influenced by ideas advocated early in the century by the Dutch magazine *De Stijl* (The Style), which had been founded in Amsterdam in 1917. (It is fair to say, though, that *De Stijl* was idealistic and spiritual in em-

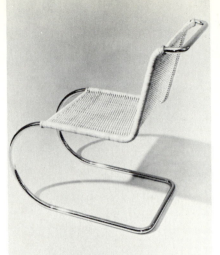

FIGURE 8A LUDWIG MIES van der ROHE. *Lounge Chair* (Barcelona chair). 1929. Chrome-plated steel bars; leather. 29″ h. Museum of Modern Art, New York

FIGURE 8B LUDWIG MIES van der ROHE. *Side Chair.* 1926. Chrome-plated steel tube; cane. 30¾″ h. Museum of Modern Art, New York

FIGURE 8C MARCEL BREUER. *Side Chair.* 1928. Chrome-plated steel tube; wood; cane. 32″ h. Museum of Modern Art, New York

FIGURE 9 CHARLES EAMES. *Lounge Chair and Ottoman.* 1956. Rosewood, leather, and steel. Private collection

phasis whereas the Bauhaus ideology was rooted in materialism.)

"Equally important" with the Bauhaus in the establishment of Nordic Modern's supremacy, says editor and writer Harold T. P. Hayes, "was the Swiss School, which in a complementary fashion began to reassess typography as a visual tool. Curlicues went out and bare-boned type came in. The absolute ultimate in no-nonsense lettering was a typeface that came to be called Helvetica [the Swiss name for Switzerland]. It remains so today. The Mobil sign . . . is a derivative."[2] Fig. 10 gives you some idea of what Nordic Modern is like. Its symbols are by no means without humor or wit. But if lightheartedness appears, it is for a reason, as part of the motif, because of the point being made and the logical nature of the information communicated.

We must stress that, although there is a certain truth to the geographic connotations of the term Nordic Modern insofar as the origins of the trend go, it would be a mistake to suppose that only Northern Europeans and North Americans cast their lot with it. What is best in Italian design (Fig. 11) is very definitely Nordic in character. More than with any locale, Nordic Modern has tended to be associated with corporations and governments, for which its reasoned-out, clean-cut symbolism provides institutional art that is coherent, attractive, and identifiable. As it happens, however, this association with corporate thought and bureaucratic sensibility is also one source of a reaction against it.

Hayes contrasted the rationality of the Nordic style with the "pungent breath" of Push Pin Studio,

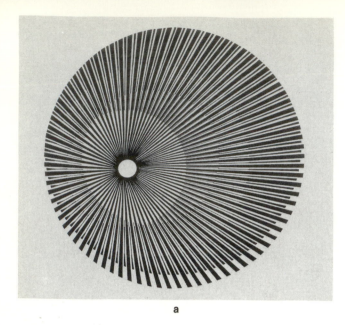

which passed over the field in the middle 1950s, gathered force in the late 50s, and swelled ''to a gale in the anarchistic 60s.''[3] Push Pin Studio, a design firm founded by Seymour Chwast and Milton Glaser, is the most prominent single representative of what has become the viable opposite extreme from Nordic Modern. That extreme is a comparatively irreverent, cluttered, and carefree manner which we

FIGURE 10 Nordic Modern graphics A: Strathmore Paper Co., Westfield, Massachusetts; B: FRED BURTON, Prize-winning proposal for new Sony logotype; C: Arressico, Logo for Duxon Educational Electronics Designs, Inc.; D: Directional Industries, Inc.; E: JOHN CELUCH, Studio 298, Inc., Edwardsville, Illinois; F: JOHN ADKINS RICHARDSON, Arressico; G: CHARLES F. WICKLER, Mongram.

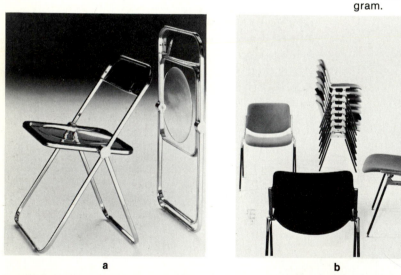

FIGURE 11 Modern Italian furniture designs. Castelli Furniture, Inc. A: GIAN CARLO PIRETTI, Plia Chair. B: GIAN CARLO PIRETTI, 106 Stacking Chair. C: GIAN CARLO PIRETTI, Axis 3000 and 4000. D: Dolmen managerial office desk grouping.

shall call *Post-Modernism.* Post-Modern is a term currently in fashion to describe a general reaction—not only in commercial and industrial design but also in the fine arts and architecture—against the monumental solemnity and impersonality of Nordic Modern and modernism generally. In graphic design, Post-Modernism deals in nostalgia, irony, exaggeration, and outright foolishness. It draws freely upon all graphic art of the past, not excluding that of the Bauhaus (see Fig. 12). As one might expect, it has been less apt to represent giant institutions than to advertise small enterprises and incidental offshoots from corporate monoliths. Posters, book and record jackets, magazine illustrations, TV advertising for the young or modish has for twenty years constituted the field of expression for the Post-Modern designers. Their work is whimsically arbitrary where Nordic Modern is functional and systematic. More and more, however, the two extremes are blurring together in synthetic blends which take advantage of both. In this book we shall try to take that kind of advantage also.

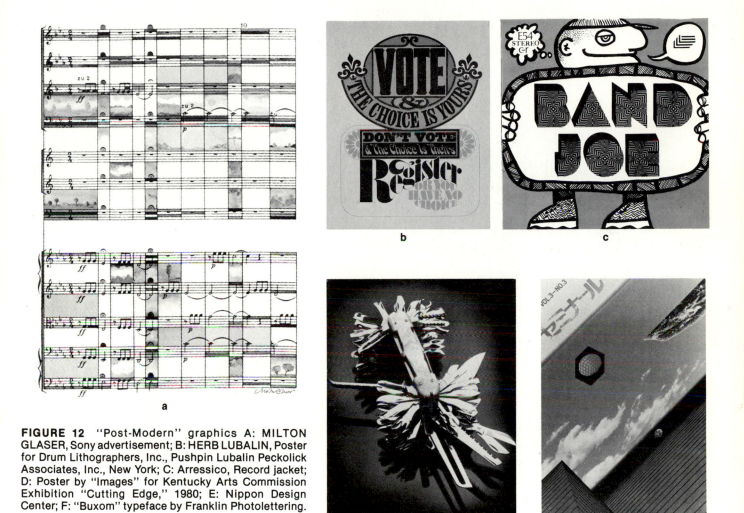

FIGURE 12 "Post-Modern" graphics A: MILTON GLASER, Sony advertisement; B: HERB LUBALIN, Poster for Drum Lithographers, Inc., Pushpin Lubalin Peckolick Associates, Inc., New York; C: Arressico, Record jacket; D: Poster by "Images" for Kentucky Arts Commission Exhibition "Cutting Edge," 1980; E: Nippon Design Center; F: "Buxom" typeface by Franklin Photolettering.

FIGURE 13A Art Nouveau design

FIGURE 13B "RI" Rock group emblem in Art Decco Style, 1981

FIGURE 13C RICK GRIFFIN. Unused dance poster featuring "The Who," 1969.

All kinds of things that a Bauhaus-influenced designer would consider retrogressive and tasteless crop up in the newer designs. *Art Decco,* for instance, a sort of *Art Nouveau* manner (Fig. 13A) adulterated by machine-made regularities and pseudo-modernist traits, was revived. Art Decco (Fig. 13B) is the very kind of thing that people who studied design during the 1940s and 1950s were taught to hate.

By 1960 and throughout the following decades Post-Modernism had more and more of an effect on the field. Even some of the older professionals began using old-fashioned graphic styles as symbolic devices. Young, radical designers who did posters for rock concerts and the like indulged themselves in a complexity and whimsy that was so antagonistic to anything ''straight'' as to be totally removed from the Nordic Modern/Post-Modernist controversy (Fig. 13C). That proved to be refreshing, too, and advertisers aiming at the enormous youth market quickly seized upon the posters as models for their ads. And so it was that the underground found its way to the top. The Post-Modernist trend had been germinating almost from the outset in the studios and printing plants of the graphic arts industry (see Figs. 14 and 15). The artists who came from Bauhaus-oriented schools had accumulated a set of attitudes that were necessarily incompatible with any business's commitment to profit for itself, its stockholders, and, ultimately, the society. Why? Because these designers had been conditioned to feel that they had a social obligation to clean up the detritus of modern industrial capitalism. It is true that usually they had little or no understanding of the background to their conviction; they had simply absorbed it from the kinds of attitudes their masters projected and the kinds of things their textbooks proclaimed. And there is no question that the intentions of the books and the teachers were noble; a good test of that was Hitler's hatred of the Bauhaus.

The Bauhaus ideology, as it was enunciated by Moholy-Nagy and his pupils, was deeply imbued with democratic socialism and had been much influenced by such thinkers as sociologist Karl Mannheim, who wished to reconstruct society along benevolent patterns through education and planning. The Germans felt, with apparently good reason at the

FIGURE 14 "In studio" graphic diversions

FIGURE 15 ROBERT RAUSCHENBERG. Retroactive I. Oil and silkscreen on canvas. Wadsworth Atheneum, Hartford, Connecticut. Gift of Susan Morse Hilles. Oil on canvas, 96 × 72″.

Individual artists and designers sometimes have rather uneasy and ambivalent relationships with the businesses and institutions for which they work. Consequently, their working spaces have been forever filled with ironic self-denigration and anti-institutional humor, frequently cast into forms featuring incongruous motifs and typefaces. Fig. 14 is a sampling of things pulled out of the files from as far back as 1950. A great deal of this "in-house" irony found its way into Pop Art (Fig. 15) and then made its way, by the back door, into the graphic designs of Push Pin Studios and other Post-Modernist operations.

time, that a well-designed world—functional, clean, and reasonable—would lead to a society that was harmonious, humane, and rational. Their main immediate goal was to reintegrate the creative artist into the reality of modern industrial society and, simultaneously, to humanize what they considered the exclusively materialistic attitudes of the business community.

Not surprisingly, businessmen and women—the people who must, after all, see to it that their enterprises operate successfully—were apt to consider the "design ethic" of the Bauhaus condescending and naively wrongheaded. At worst they considered it sinister. Ayn Rand, in her novel *The Fountainhead* (1943), attacked the collectivist mentality of altruistic designers by confronting their ethic with the self-centeredness of an untraditional architect who despises the masses and designs buildings to satisfy his

personal integrity and sense of form. The author sees the functionalism of her hero's architecture as being an expression of the "functionalism" of the business world which runs according to sheer pragmatism. No matter how simpleminded Ms. Rand's logic seems, and whatever else one may think of capitalism, it must be admitted that there is a crude kind of reasonableness to the claim that competition for profit provides the most clearly *objective* system of assessing products. Here, of course, is where the designer comes into the picture. Everyone knows that an effective advertising campaign can overcome qualitative deficiencies, that "styling" an appliance in certain ways can make an inferior machine look more efficient and functionally more sound than a competing unit which works a lot better. Clever packaging of a bad nutritional buy in breakfast cereal can so enhance the sales appeal of the brand that it becomes the mainstay of childhood diets all over the country.

It is because of the conflict of businesslike hard-headedness (or ''ruthlessness'') with Bauhaus-like humanitarianism (or ''do-gooderism'') that design educators and textbooks often strive to separate ''graphic design'' from the ''hard-sell conventions of 'commercial art' with its willingness to subordinate itself to the marketplace.'' But the problem is *real,* not just semantic. What can a designer do when a client or superior wants something worked up to sell a product that is dangerous to the society? Argue against producing the thing? Good luck! Refuse the commission? Quit the agency? Those are genuine alternatives, surely, but rarely are things so flagrantly awful or firms and government bureaucracies motivated by such apparent evil that we ordinary human beings will give up security or income for the sake of some anonymous consumer. Milton Glaser has said that whether one ought to help a client whose objectives are inimical to society is ''a difficult question for all designers'' and that it is a problem he has ''with advertising.'' The easiest course is simply to shift the responsibility for society's welfare to the client and let him or her worry about the impact of the product.

That is what will usually happen.

Whether designers should take it upon themselves to solve the problems of an ever noisier, filthier, and more illiterate society is a serious ethical question. But the fact is that it is not feasible for most of them to do so.

Contemporary design echoes the past and draws upon the present; its symbols and its shapes can be beneficial to society or not, as the society wills. A designer can work for profit only, for social justice, or for the gratitude of a small avant-garde clientele. Advertising agencies, government bureaus, publishers and charitable institutions, universities and museums all employ designers. The world of commerce compensates them very well with comparatively healthy salaries and commissions. Public service affords the comforts of security and the self-satisfaction of helping out one's fellows. It is always important in life to understand just who you are and what you really want to do as a member of society. In the case of design, it is most important first to learn the fundamentals with which one must work.

And so, we begin.

notes for chapter 1

[1] Milton Glaser, quoted in *The Complete Guide to Illustration and Design,* ed. Terence Dalley (London, Chartwell Books, Inc., 1980), p. 104.

[2] Harold T. P. Hayes, ''The Push Pin Conspiracy,'' *The New York Times Magazine,* March 6, 1977, p. 20.

[3] *Ibid.*

part 2

systems

chapter 2

grids, mazes, and modules

A "system" is simply an orderly way of doing something. In art and design it might be a way of analyzing works or it might be a way of organizing shapes and colors into some sort of unity. Thus, art that seems to result from the application of a specific procedure, from repeated use of a pattern or set of patterns, or from adherence to a body of rules is sometimes referred to as "systemic art." An example of such a style could be the grid paintings (Fig. 16) of Agnes Martin (b. 1912) in which not quite horizontal and nearly vertical lines create a sort of veil, delicate, yet full of unexpected power when seen at first hand. Critic Lawrence Alloway has said of Martin's work that she "can fill the house with a whisper."

A grid is a system of fixed horizontal and vertical divisions and, while Ms. Martin's works reveal the charm of slight deviations from regularity, grids are among the most adaptable and universally applicable of all systems. Anything, no matter how irregular, can be conceived of in terms of what coordinate geometers would call X (horizontal) and Y (vertical) axes. Thus, a *serigraph* print (Fig. 17A) designed by Martin long before her gridwork period can be described in terms of a grid superimposed upon it (Fig. 17B). Indeed, halftone reproduction, which breaks subtle photographic grays down into black and white for printing, does precisely this. The tiny dots that result from the procedure as it has been applied in the book you are reading are too fine to reveal that, since there are over 100 of them per square inch. However, if we use a very coarse halftone, having 40 lines per square inch, and then enlarge it, the effect becomes evident (Fig. 18). And, if you look at Fig. 18 from ten or twelve feet away, the grays will return to the reproduction. The technique used to create halftones need not detain us just now; the point is that the effect results from imposing upon a photograph a very fine gridwork of regularly spaced lines.

The mechanical regularity of a grid can also be used to reveal more complicated relationships within a work of art. Michelangelo's oil painting *The Holy Family* (Fig. 19A) is a useful example because it not only is a work of enormous skill but is also a circular canvas (properly called a *tondo*) and so might seem resistant to a system of angular relations. But Fig. 19B gives immediate evidence that this is not the case.

FIGURE 16 AGNES MARTIN. *The Tree.* 1964. Oil and pencil on canvas, 6 × 6'. Museum of Modern Art, New York, Larry Aldrich Foundation Fund.

FIGURE 17A AGNES MARTIN. *Untitled.* 1952. Serigraph, 9 × 12″. Private collection

FIGURE 17B Grid superimposed on serigraph by Agnes Martin

FIGURE 18 Detail of serigraph in 40-line halftone, magnified

FIGURE 19 MICHELANGELO. *The Holy Family,* called the *Doni Tondo.* c. 1503. Oil on panel, diameter 47 ¼″. Uffizi Gallery, Florence

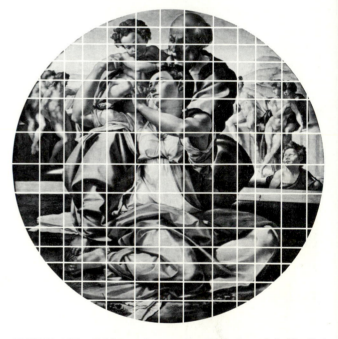

FIGURE 19B Grid superimposed on Michelangelo's *The Holy Family*

So many significant edges and prominences fall on crossings or within rectangles here that it is tempting to argue that a grid is the *basis* of this composition. That argument would, however, be mistaken. There are too many relationships it cannot explain. Among these are the pathways formed by linking up edges of shadows and outlines that do not fit into the grid.

Still, if our grid does not fully account for the design of the painting, it does at least evoke one aspect of Michelangelo's organization.

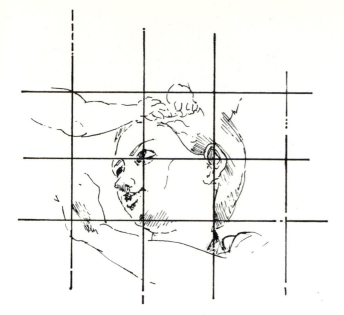

Something else about the grid as it has been employed in Fig. 19B: If you wished to copy the painting fairly accurately but on a larger scale, it would be a simple matter to draw the same number of lines twice as far apart on a circle with a diameter twice that of the reproduction. Or four times. Or twenty. Then, by carefully redrawing the outlines that occur in each box (as in Fig. 20), you would be able to approximate the proportions of the reproduction while at the same time magnifying it. Artists do sometimes employ exactly this technique for expanding the size of a sketch so that it can serve as a basis for a finished work. A famous example in which the cells can still be seen is *Third Class Carriage* (Fig. 21) by Honoré Daumier (1808–1879).

Grids can be used in many different ways. Since their pattern need not be based on identical spaces but can depend upon *any* regulated distancing of vertical, horizontal, or diagonal elements, many possi-

FIGURE 20   Enlargement by means of a grid pattern

To enlarge a picture the way Daumier evidently did his *Third Class Carriage,* you must be sure that the cells on the sketching paper and the larger ones on the canvas are in exactly the same proportion. You cannot squeeze what is on a 3″ × 5″ piece of paper onto a canvas four feet square because the two rectangles are not in proportion. But, assuming that you have a canvas that is 3′ × 5′ (or 6′ × 10′, or 12′ × 20′, or anything in the same proportion), here's what to do: Divide the sketch in half vertically, then into squares, then eighths, and so on, until you have as many sections as seem useful. (This is easy to do without a ruler: for example, take a long strip of paper and fold it in half; then halve the half, halve that half, and so on.) Then divide up the sketch the same way horizontally. Once this has been done, carry the same procedure over to the canvas, being sure to create the same number of boxes.

FIGURE 21   HONORE DAUMIER. *Third Class Carriage.* c. 1862. Oil on canvas. 25¾ × 35½″. Metropolitan Museum of Art, New York, H.O. Havemeyer Collection

FIGURE 22 Traditional building facade with a grid imposed upon it.

FIGURE 23 Modular systems in house design A: Derived from MIES van der ROHE, Edith Farnsworth House, Plano, Illinois; B: Derived from JOHN ADKINS RICHARDSON, Ravine House, Edwardsville, Illinois

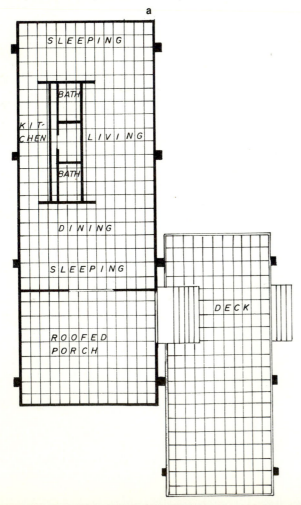

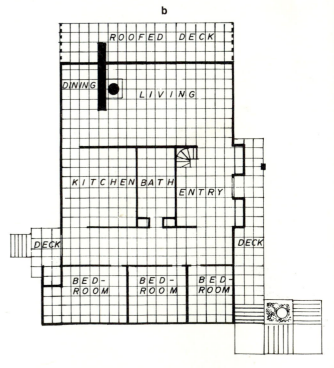

bilities are open to designers using them (see Figs. 22, 23, 24, 25, and 26).

In *Numbers in Color* (Colorplate 1) Jasper Johns (b. 1930) restricted himself to a set of identical cells and numbers that are regulated in terms of appearance and dimension by the conventions of lettering. That is, normally, every time a given numeral in a given lettering style appears at a given size it looks exactly the same. And here, the ten ordinals (0–9) are arranged conventionally. Reading from left to right, they are in the normal sequence. From top to bottom they read as they do from left to right. Diagonally, from top left to bottom right, they fall in rows alternatively odd or even; from top right to bottom left a given row has the same number. Everything about this scheme and the elements from which it has been conceived is intellectual, regimented, calculated. But look at what Johns has done to the table of numbers! No two are painted in the same way. The colors vary and the way the pigment has been applied does too. Each letter form has been changed and given an individual character by the artist so that what must have looked very unimaginative and routine when first laid out on the canvas has been turned into a delight-

Architects sometimes use grid relationships in designing buildings. In older structures the application of the coordinates is usually very obvious (Fig. 22), but some contemporary architects apply a similar set of units or *modules* when they articulate floor plans and facades (Fig. 23). A module is merely a standard unit of measurement employed throughout a building or other structure. Sixteen inches is frequently used in frame construction because that is the distance that studs (the vertical posts in a wall or partition) are usually spaced. In ancient times the diameter or radius of the base of a column customarily served as a module from which all other dimensions were derived.

FIGURE 24 VICTOR VASARELY. *Vega.* 1957. 27 × 51″. Collection of the artist.

This is an Op Art design which depends for its effect upon the progressive expansion of the grid from a central point. It has, as you can see, illusionistic properties; the surface seems to be warped in space.

FIGURE 25 FLOYD COLEMAN. *Neo-African #12.* 1977. Mixed media on paper, 9½ × 9½″. Private collection.

Coleman's *Neo-African #12* (Fig. 25) is one of a series using a broad grid as the basis of the designs. Within the boxes, textural variations are achieved through the building up of layers of strokes with pencils, pens, felt markers, crayons, charcoal sticks, and pastels. The artist's inspiration was African fabric (Fig. 26).

FIGURE 26 African fabric (Yoruba)

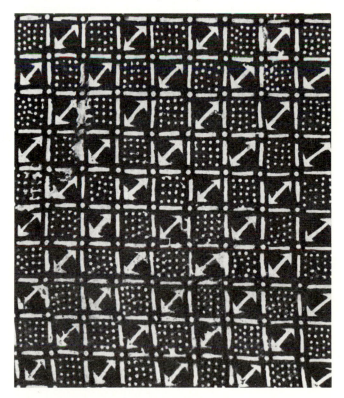

ful exercise in variation. The numerals are not as legible as they must have been to start with, but they are much more intriguing. Still, the underlying consistency of the lettering and the fundamental regularity of the grid provided a unified structure which was able to tolerate radical improvisations.

Fig. 27 shows four examples of grid designs we've dubbed "supergraphs," done by university students in a first-year course. Details from the same pieces appear in color as Colorplate 2. It would be possible, surely, to translate the "supergraphs" into large-scale paintings, tapestries, or prints. A somewhat similar procedure must have been employed to create the serigraph print in Fig. 28.

Artist Robert Kirschbaum (b. 1949) has done drawings and prints (Fig. 29) which resemble nothing so much as the diagrams of buildings but which have, also, a kind of occult mystery. The black squares, which might be taken to stand for the bases

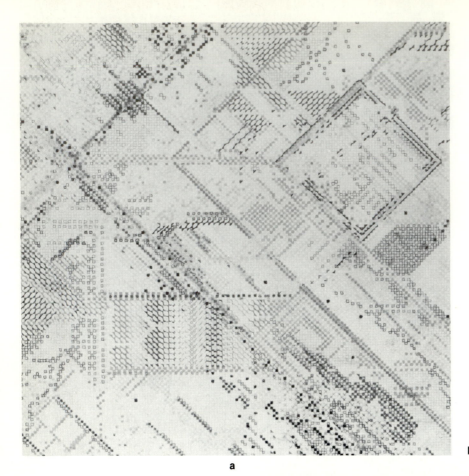

a

FIGURE 27 Four "Supergraphs"

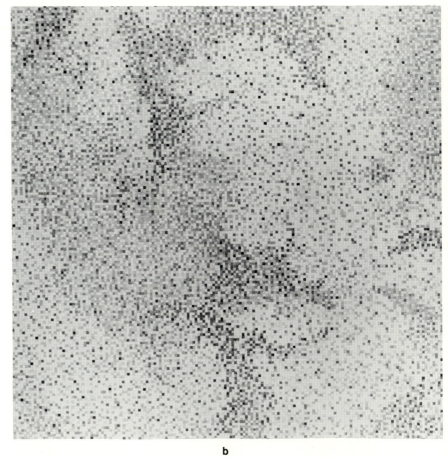

b

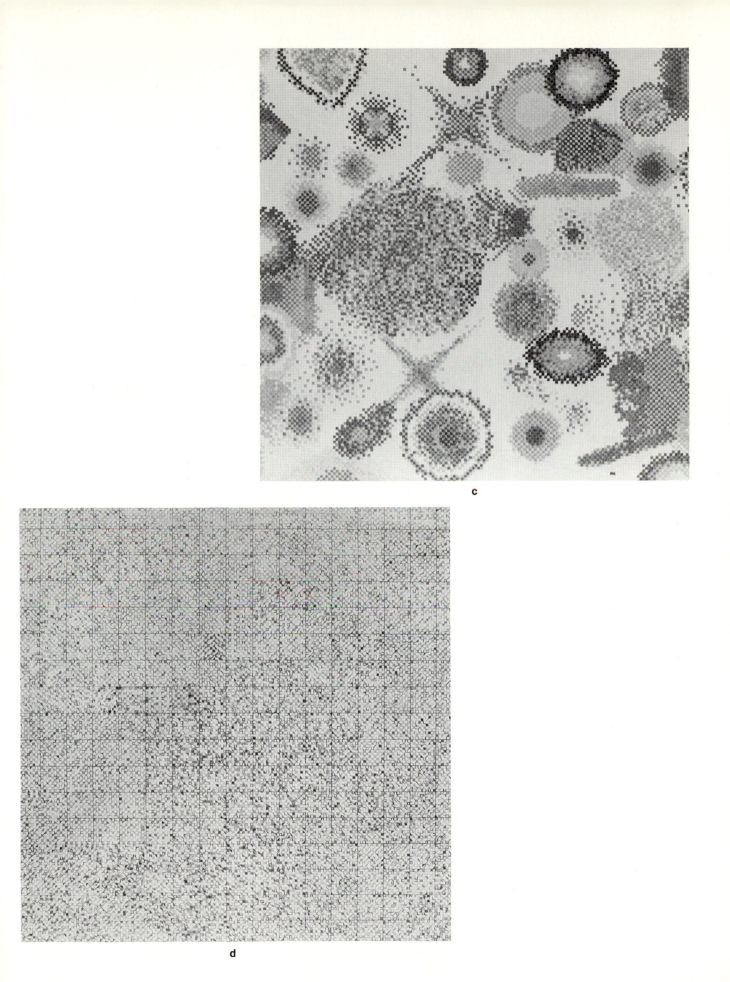

c

d

FIGURE 28 LODEWIJK. *Untitled.* 1971. Serigraph, 19½ × 19½". Collection Southern Illinois University at Edwardsville.

of columns, also hang so sharply dark against the gray tracery of lines that they seem like windows or doorways; they pierce the pictures with darkness. It is as if they opened onto a backdrop of eternity.

One of the things a regulated grid lends itself to is *anamorphic distortion.* Look at Fig. 30A. Compare it with 30B. The square cells have been "stretched," as it were, so that they are three times as long as they are high and the elements of the head inscribed within them have been similarly altered. If, however, you turn the page of the book at an abrupt angle to your line of sight, Fig. 30B will assume the same appearance as Fig. 30A seen straight on. This is an instance of anamorphic distortion. The root word is from the Greek *anamorphosis,* meaning "to form anew."

The art of the phenomenal German Swiss artist Hans Holbein the Younger (1497/98–1543) contains one marvelous example of anamorphic imagery and is affiliated with another, the anamorphic version of his portrait of England's Prince Edward VI (Fig. 31). But his own work contains another famous example. In *The French Ambassadors* (Fig. 32) Holbein has portrayed two emissaries from France to the court of Henry VIII, father of Edward VI and the sovereign for whom Holbein had become official artist.

The anamorphic skull on the floor, once revealed (see Fig. 33), gives the delicate constructions,

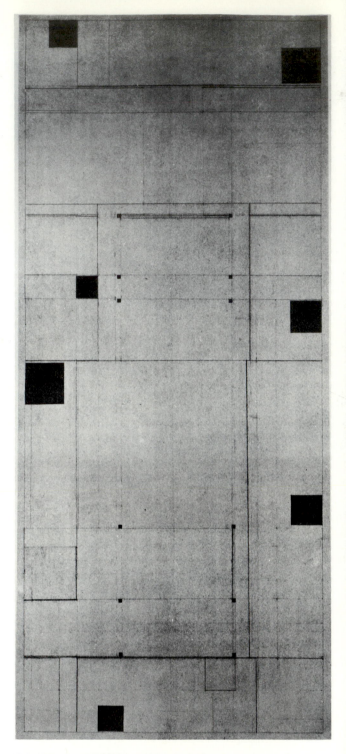

FIGURE 29 ROBERT KIRSCHBAUM. *Temple of Solomon.* (State 2). 1978. Diazo print, 48 × 20". Private collection.

globes, and musical instruments on the table an additional meaning. The presence of the skull turns the picture from a mere double portrait into a *vanitas,* a painting emphasizing the ephemerality of human life and pride. Too, the man on our left, Jean de

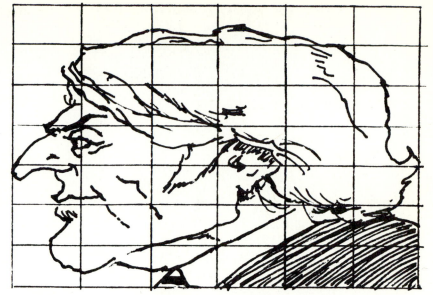

FIGURE 30A Drawing of a human head within a grid

FIGURE 30B Anamorphic distortion of Fig. 30A in the horizontal plane

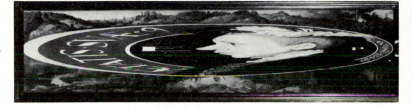

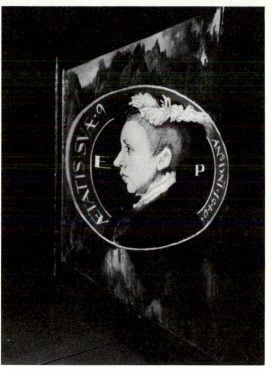

FIGURE 31 WILLIAM SCROTS. *Anamorphic Portrait of Edward VI, from front and side.* 1546. Oil on panel, 16¾ × 63″. (after Holbein painting of 1543). National Portrait Gallery, London

Dinteville, wears a skull-embellished medallion on his cap. The ambassadors are showing off Gallic culture but at the same time giving, through Holbein, a modest gesture which says, in effect, that all is the work of erring mortals. In other words, criticism is turned away by anticipating it in a very witty fashion. This is not to say that there might not be other motives for the inclusion of skulls.

When a grid is sufficiently liberated from the vertical and horizontal, it becomes a random criss-crossing of lines—a scribble. While we tend to think of scribbling as being without redeeming artistic value, it is actually more fluid and coherent than it is chaotic. In the work of a Swiss artist born centuries after Holbein, Paul Klee (1879–1940), one can discover all sorts of whimsical diversions (see Fig. 34), some of which resemble shapes that can be evoked from scribbled patterns by employing a technique described in Fig. 48 and illustrating Exercise 2–5 at the end of this chapter.

FIGURE 32 HANS HOLBEIN. *The French Ambassadors.* 1533. Oil on panel, 81¼ × 82¼". National Gallery, London

FIGURE 33 Detail of anamorphic portion of *The French Ambassadors* from the proper angle to reveal the skull

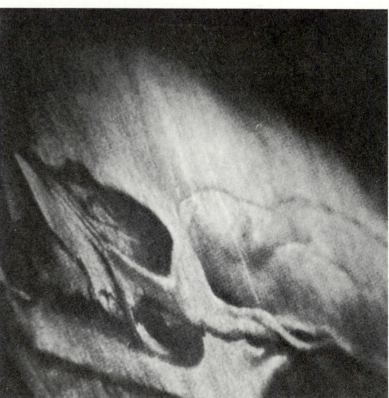

Hans Holbein's ability to render the human presence with an authenticity that sometimes excels yet-to-be-invented photography is uncanny. In some ways this artist's realism has never been surpassed. The versimilitude of his portraiture—recognized in his own time and therefore verified for ours—is no more astonishing than his capacity as a still life artist. In this work the two-tiered table behind the Frenchmen is filled with fragile instruments which indicate that the men have an interest in the life of the mind. Resting on the floor, though, is a mysterious, elongated object. When looked at from just the right vantage point, this queer shape takes on the semblance of a human skull (see Fig. 33).

FIGURE 34 PAUL KLEE. *Land-scape with Yellow Birds.* 1923. Watercolor and oil on colored sheet, 14 × 17⅜". Private collection, Basle

Klee's paintings disclose a delightful world full of lively and charming things, such as the bird who is taking an idle stroll along the bottom of a cloud. Such fauna are more or less routine images for Klee, but the flora here bloom with a robust inventiveness that is like unaffected nature more than it is like artificial craft. These shapes look, in fact, like the forms we shall presently evoke from mere scribbling!

FIGURE 35 JACKSON POLLOCK. *Reflection of the Big Dipper.* 1946. Oil on canvas, 43⅝ × 36¼". Stedelijk Museum, Amsterdam

Jackson Pollock could, as he was at pains to point out, control the flow of poured paints, as anyone who has poured syrup over pancakes or chocolate onto ice cream will know. It is possible to fatten a line by pouring slowly or make it thinner by going faster. One can direct the general path of fluid quite accurately, so long as the paint or syrup is viscous enough not to run at random all over a surface. Pollock followed a very deliberate procedure, laying in a background pattern to establish large, dominant movements that would decide the composition of the picture. Then, with another color, he created a series of subordinate movements, usually more intricate ones. As the work progressed, the artist embellished and articulated a maze of lines until, finally, he had a tremendously complicated structure.

Jackson Pollock (1912–1956) was a painter whose "scribbling" rose to the level of the highest art. In 1947 he abandoned brushes and created a style of painting (Fig. 35) which depended on a technique of pouring paint from cans directly onto canvas tacked onto the floor. This procedure, which Pollock employed until 1953, may seem to be altogether mad but it produced some very exciting and beautifully rich paintings, usually on a very large scale.

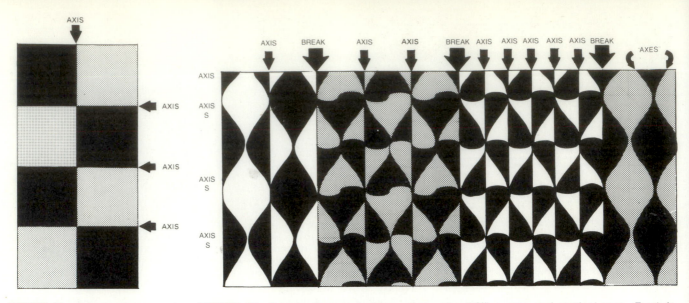

FIGURE 36 Counterchange pattern of squares with the vertical and horizontal axes pointed out

FIGURE 37 Counterchange designs using ogive ("S") curves and vertical axes. Far left: typical ogive "onion" pattern split by axes. Left center: same principle, using onions ornamented with a horizontal S-curve. Right center: additional vertical axes through all onions. Far right: axes following ogive contours.

FIGURE 38 Counterchange design having both vertical and horizontal axes, like a checkerboard.

A design device very closely associated with the grid system is called *counterchange*. The simplest version of this is a checkerboard. Consider each line on the checkerboard as an axis and examine Fig. 36 in that light. You will realize at once that the axes are flanked by black and gray squares which alternate. Counterchange is nothing more than the reversal of shapes and/or colors along an axis. Usually, the axes are parallel in one direction only, as in Fig. 37, and not in both the vertical and horizontal dimensions, as the checkerboard and Fig. 38 are. Most commonly, the axis in a counterchange design is a straight line, but it is quite possible to make it curvilinear, jagged, or whatever, and to vary the size and angle of the component shapes along the axis (see Fig. 39).

Obviously, the counterchange design is most commonly encountered in the kind of overall patterns seen in wall coverings (Fig. 40) and textile designs

FIGURE 39 Counterchange patterns featuring various irregularities

FIGURE 40A Wallpaper pattern design

FIGURE 40B Tiles from the Alhambra, Granada, Spain

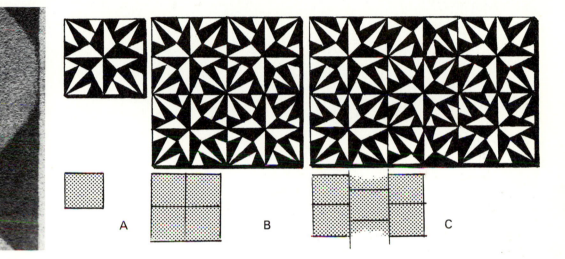

A   B   C

FIGURE 41 Seven Seas towel in a counterchange design by Cannon Mills Company. Detail.

FIGURE 42 Modular unit and pattern repetitions. The unit (A) is a very old geometric star familiar to inlay craftsmen. In block arrangements (B), it produces also a rosette wherever four units join. In a staggered arrangement (C), the stars are framed by accordian-like swirls.

FIGURE 43 M.C. ESCHER. *Sky and Water I.* 1938. Woodcut. 17½ × 17½". Private collection

(Fig. 41). Such designs consist of a modular unit that is repeated over and over to give the impression of an endlessly renewing pattern (Fig. 42). Sometimes such units are repeated in a simple grid pattern; at others they are staggered or alternated with other units to make more complicated surfaces.

A sophisticated version of counterchange appears in Fig. 43, one of M. C. Escher's disconcerting illusions in which the axes are staggered and the contrast not merely tonal but also representational. Variations have been rung on this kind of interlocking identity of forms in trademarks and in fine art as well (Fig. 44).

FIGURE 44 MICHAEL BUESE. *Counterchange #2.* 1969. Serigraph, 17 × 17″, Private collection

exercises for chapter 2

EXERCISE 2–1: Drawing Grids

Materials: *White drawing paper, 12″ × 18″*
Tools: *Yardstick or long ruler*
HB pencil
T-square
Triangle

One really needn't make one's own graph paper, of course, but the following exercise is valuable because it provides a quick and easy technique for dividing any space into any number of equal spaces.

Let us say that you want to divide a square into 64 uniform cells. This means that you must divide it into eight equal segments from left to right and eight equal segments from top to bottom. Now, if the square is 8 inches on all sides, or a number divisible by 8, the solution is immediately apparent. However, if the measurement is an odd one, like 5 8/15 inches, then division—even with a calculator—is not too neatly decided. There's really nothing to it, however.

Take the ruler and move it around until you find a convenient number of inches totaling eight. Zero and eight are the obvious demarcations in this case. Match the zero (that is, the end of the ruler) to the top line and the 8-inch mark to the bottom (Fig. 45). Then mark off each inch and draw parallel lines

FIGURE 45 The division of a rectangle of any size into equal parts

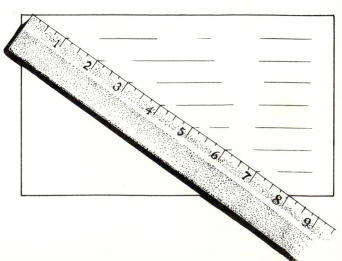

FIGURE 46 An irregular op-art grid

through these points with a T-square. Use the same technique to divide the square vertically and draw the vertical lines with a triangle.

If your square had been too large for zero and eight to reach, you could have doubled or tripled the measure. (That is, use zero and sixteen to mark off every 2 inches instead of every one. Or use zero and 24, marking off 3-inch intervals.) Once the horizontals have been drawn, establishing the verticals is even easier because you just have to draw a diagonal line from one corner to the other and put verticals in wherever the diagonal intersects an upright. It should be obvious, too, that what can be done in a square will work also on a rectangle. (Of course, if you want the cells to be square, you will have to work within the squared end of the rectangle and then carry the same measurements out along the remaining end.)

This technique always works, no matter what the distance is. You can divide with perfect accuracy without any measuring. Also, the divisions do not have to be equal because you can mark off each differently—say, 1 inch, then 2, then 3, then 1, or whatever. See if you can produce an interesting variation by employing this technique. Fig. 46 is one example of such a variable grid. The painting in Fig. 24 might well have evolved from just such a procedure. Too, a grid need not be made up of lines parallel to the edges of the rectangle. The lines might be diagonal or might not be at right angles to one another. Certainly, the cells formed by intersecting lines may be rectangular instead of square.

EXERCISE 2–2: Supergraphs

Materials: One sheet of white graph paper, 17 1/2" × 21", ten squares to the inch
Tools: Colored fiber-tip pens, colored pencils, color watercolor pencils, or color crayons in any combination or by themselves

Create a design, using the entire sheet of paper. A minimum of one half the squares must be filled. You may use any of the following techniques, subject to the restrictions listed after them.

The squares may:

1. be completely filled in and consist of geometric units of different hues (such as a triangle of red and one of yellow or a half square of blue and a half square of orange).
2. be filled with straight line elements.
3. contain free-form linear elements.

But:

1. the same color or linear device may not occur adjacent to itself for more than one square vertically or horizontally, although it may diagonally.
2. no recognizable subject may be used and no letter form may appear.
3. when the design is finished, no overall pattern should be evident.

Fig. 27 shows four student solutions to this problem, and Colorplate 2 contains details from the same solutions.

EXERCISE 2–3: Anamorphic Distortion

Materials: White drawing paper
Tools: Yardstick or long ruler
HB pencil
Pen and ink
T-square
Triangle

Select a simple drawing or design that you have done and redraw it in a grid made up of squares, as in Fig. 47A. Then, keeping the lines at right angles, draw another grid with cells elongated so that the vertical and horizontal magnitudes are unequal and distort the drawing correspondingly, as in Fig. 47B, where we have used only the first column of the abstract design for our *very* elongated example. Do the same a second time, but retain the proportions and merely change the angle of the horizontals with respect to the verticals (as in Fig. 47C). Another possibility is to make the horizontals into diagonals that converge to a single point off to the left or right and diminish the distance between the verticals by some regular percentage as they approach the point of convergence (as in Fig. 47D). This last produces an effect very similar to perspective distortion. Vastly more patterns are possible, including those involving curved lines and staggered cells.

Anamorphic images are antique, but they still have many uses today and, like many other effects, can be approximated with photography. Since not all anamorphic effects are easily simulated with a camera, however, it is wise to gain understanding of the device through direct experience.

FIGURE 47A Drawings within grids;

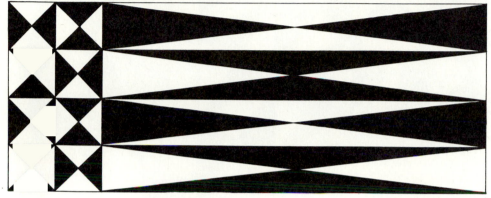

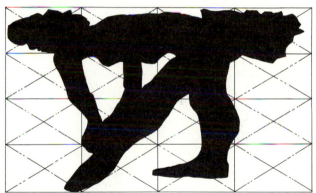

B: Fig. 47A distorted anamorphically at right angles and unequal magnitudes;

C: Fig. 47A distorted anamorphically on a diagonal with magnitudes equal;

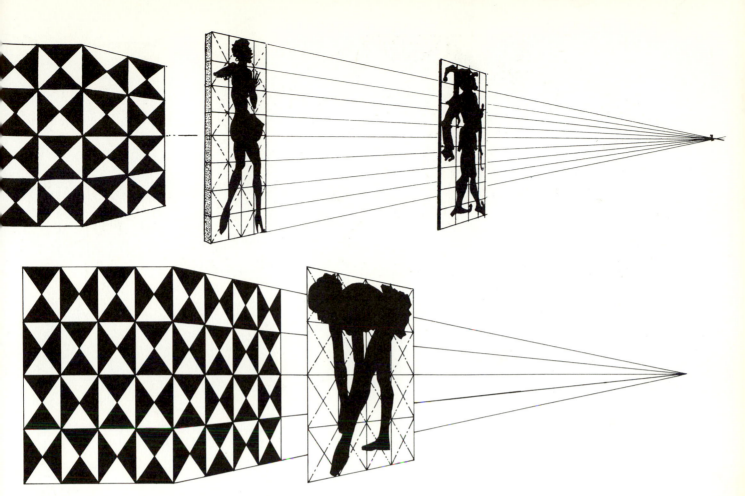

D: Fig. 47A distorted anamorphically with converging horizontals and diminishing magnitudes

EXERCISE 2-4: Scribble Designs

Materials: *White drawing paper*
Tools: *Black fiber-tip pen and other drawing implements*

Take a sheet of paper and a pen and start scribbling. Keep scribbling until you have something like Fig. 48A. Then, fill in some of the zones (see Figs. 48B and 48C). When you have some forms that you think you may be able to use, cut them out or trace them. Then use them to make up some sort of image. Fig. 48D contains several things students of various predilections have done. They are not Paul Klees (Fig. 34), certainly, but they do have their own kind of spontaneous integrity.

EXERCISE 2-5: Counterchange Pattern

Materials and tools: *Of your choice*

Using the principle of counterchange, devise a unit approximately 6 inches square that can be used to produce an overall pattern, after the procedure used in Fig. 41.

FIGURE 48A Scribbling;

B: A scribble with some elements filled in;

C: A scribble with some elements filled in;

D: Scribble-inspired designs by students

fields, forms, and fragments

A grid could be taken apart or cropped to produce something like Fig. 49, and the result is similar, perhaps, to a painting by Piet Mondrian (1872–1944). But, in point of fact, a Mondrian (Fig. 50) is much subtler than this makes it seem. The divisions of the painting are not uniform and even the thickness of the stripes varies slightly. The artist was extremely serious about his work; he felt that by reducing the elements of painting down to such fundamental contrasts as vertical and horizontal, light and dark, and neutral and primary colors he approached a harmony and orderliness that expressed the essential nature of reality—not the ordinary, prosaic reality of everyday vision, but an unchanging reality or truth which lies behind all things, an eternal reality that transcends any time, place, or universe. Of this notion, the artist said that we are, indeed, "able to see unity in natural things. However, there it appears under a veil."[1] His works, he felt, might reveal the unity symbolically and therefore more clearly.

Mondrian's comment appeared in the Dutch journal *De Stijl,* referred to earlier, for he was a member of the Nordic Modern group, which sought to purify art and functionalize industrial design. His style has its own name; it is called *Neo-Plasticism.*

Despite the rather pretentious goals the artist proclaimed for himself, his style has had an enduring influence, partly because his conception of stripes spanning a canvas in an asymmetrical arrangement is very like the realistic Impressionism from which modern art evolved (see Figs. 51–54).

Generally speaking, modern art has tended to be more like the Degas and the Mondrian than like the Albers. The artist Hans Hofmann (1880–1966) did a number of paintings during the last decade of his life featuring large, boldly colored rectangles in heavily applied oil paint (Colorplate 3). Like the Mondrians, they are usually asymmetrical and the relationship of one zone to another is very carefully worked out. Hofmann liked to talk about the "push/pull" of forms and colors. If you study *The Golden Wall,* you may be intrigued by the way the units seem to lie ahead or behind one another in space. Often, the color produces an effect quite different from what the positioning of the rectangles would by itself convey. This visual game is part of the meaning of the painting. The color combinations may at first strike you as being garish, but a little patience should prove that gardens rarely blossom with so rich a display. Seeming clashes are really means of driving forms forward or backward; it is part of Hofmann's game. *The Golden Wall* does not represent a wall; it *is* a wall, a "golden" one in the sense that it is richly surfaced. The spartan neatness of the Mondrians may hold greater appeal for you, but Hofmann's bright puzzle is as cool, in its way, as is Mondrian's division of white space with black stripes.

The white zones in the Mondrians are very subtly spaced. The artist spent hours, days, even weeks arriving at them. Still, it may not be easy to appreciate the delicacy of his adjustments. Take Figs. 51 and 53 and try thinking of the white spaces as having been laid out on a black background. The exquisite proportioning of the various elements may be more apparent when seen in this light than when looked at in the expected fashion. Somewhat the

FIGURE 49 Asymmetrical section of a grid

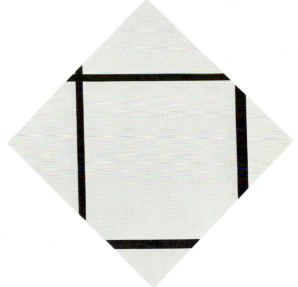

FIGURE 50 PIET MONDRIAN. *Composition in Black, White, and Red.* 1936. Oil on canvas, $40\frac{1}{4} \times 41''$. The Museum of Modern Art, New York. Gift of the Advisory Committee

FIGURE 51 EDGAR DEGAS. *At the Milliner's.* 1882. Pastel, $30 \times 34''$. Metropolitan Museum of Art, New York (H.O. Havemeyer Collection)

FIGURE 52 PIET MONDRIAN. *Painting I (Composition in White and Black).* 1926. Oil on canvas, $44\frac{3}{4} \times 44''$ (diagonal measurements). The Museum of Modern Art, New York. Katherine S. Dreier Bequest.

Art historian Meyer Schapiro has used *At the Milliner's* (Fig. 51) by Edgar Degas (1834–1917) to indicate a number of the similarities between the styles of early and later moderns. The most important of these is the way in which primary elements run off the edges of the pictures so that we seem to be viewing one of many possible segments of space. That is, the Degas has a "snapshot" quality, as if it were a random view out of many possible views of the hat shop. The lines in the Mondrians give the impression of extending beyond the canvas indefinitely. That is very apparent in Fig. 52, where it is nearly impossible to avoid feeling that the black stripes extend into the space beyond and that the painting is a representation of a section of an infinite universe full of such rods.

FIGURE 53 MICHELANGELO. *The Holy Family,* called the *Doni Tondo.* c. 1503. Oil on panel, diameter 47¼". Uffizi Gallery, Florence

FIGURE 54 JOSEPH ALBERS. *Homage to the Square: Apparition.* 1959. Oil on board, 47½ × 47½".Solomon R. Guggenheim Museum, New York

Michelangelo's *Holy Family* (Fig. 53) represents a different kind of formal conception than Figs. 51 and 52. In composition it is a good deal more like the abstraction by Joseph Albers (1888–1976). (See Fig. 54.) In the Michelangelo and the Albers things do, of course, extend to the very edge of the canvas. However, the Madonna, Christchild, and Joseph of the former are so arranged and positioned as to make the rather incongruous nude athletes and their setting altogether subordinate to the central group. Michelangelo's is no haphazard piece of the world; it is a selected space and the forms within it have been arranged to emphasize the family group. Similarly, Albers does real homage to the central forms in his painting; those squares within the square canvas are given enormous dominance over any other aspect of reality. The most obvious difference between the Michelangelo or Albers and the Degas or Mondrian is that the former are relatively symmetrical and the latter asymmetrical. The former set of pictures is more formal and the shapes they contain far more explicitly fixed in space.

same procedure may reveal qualities in the Hofmann that are not readily apparent at first glance. It is usually instructive to examine *any* picture in terms of the shapes the background takes from the shapes that seem to overlap it (see Fig. 55).

In Colorplate 4 the forms all go together whether they are positive or negative. The artist, Stuart Davis (1894–1964), created in this work what might be called a "family of forms." They are related to one another. Granted, the star in the lower left might not seem to be a legitimate member of the same family as the spots on the upper right were we to encounter the star and the spots together without anything else. However, when we see them with all of their relatives—the jagged ones, the lumpy ones, and the irregularly hacked-up ones—then the star and the spots show their family resemblance. They are merely extremes within a group. It would be easy to detect an outsider. A perfect square wouldn't belong here. Neither would a spot of pale yellow-green. Nor

would lettering that is more geometrically regular or smoothly fluid than the artist's signature. (This is not to say, however, that there might not be times when such lettering would be appropriate in this context. You might want the lettering to stand out in contrast to its setting. In that case, the lettering would not be a member of the same family of forms as the one the artist has chosen.)

Colonial Cubism is an oil painting on canvas but it looks as though it might have been created by gluing scraps of colored paper down onto a red-orange background. That's entirely possible. Stuart Davis might very well have begun in just such a way and then rendered what he had pasted up in oil paint on a larger scale. And why not create designs with pieces of colored paper? It's a good way to try out shape and color relationships, and designers do it frequently.

People have been mounting and pasting pictures together into valentines and the like for a long time, but it was not until 1912 that artists made much

a   b

FIGURE 55 A: Background shapes abstracted from Figure 51; B: Background shapes abstracted from Figure 53

In the Michelangelo the background fragments "go together," that is, they look alike (see Figure 55A). So do those in the Degas (See Fig. 55B). However, the Degas shapes are not like those in the Michelangelo. Too, in each case, background shapes (or *negative areas*) have a certain power and authority. They seem to be more than mere accidents; they appear to have been given a specific character deliberately. Notice, for instance, the way the curve of the hat the modiste offers is matched by the edge of the buyer's left arm and the way the right elbow of the latter echoes the same hat. These similarities give the background forms a quality of relatedness that happenstance realism could not.

FIGURE 56 PABLO PICASSO. *La Suze.* 1912–13. Pasted paper with charcoal, 25¾ × 19¾". Washington University Gallery of Art, St. Louis

of this activity. Then the Cubist painters Pablo Picasso (1881–1973) and Georges Braque (1882–1963) created what they called *papier collés* (Fig. 56), the first examples of this sort of thing being undertaken as serious studies in composition. The things incorporated into the example are not photographic; they are actual newspaper, an aperitif label, truly embossed wallpaper, and so forth.

The generic term for art work made by gluing various elements together to create a design is *collage,* which means "glued matter" in French. Serious artists have also created the related works called *montages,* combinations of pictures into a single picture. The term *montage* derives from French for "mounting."[2] (A *photomontage* is, of course, a montage of photographs. The first was made in 1839 by William Henry Fox Talbot.) Most often the montages created by fine artists have resembled those of the *Dadaists* and *Surrealists,* who evoked dreamlike imagery by appealing to unreason and by combining things in unlikely and incongruous ways (see Figs. 57 and 58). Black artist Romare Bearden (b. 1914) has combined photographs of African artifacts and American habits and habitats to create a series of aesthetic documents (Fig. 59) on the history and circumstances of Afro-Americans, combining a comment on the cultural

FIGURE 57 HANNAH HOCH. *Cut with the Kitchen Knife.* 1919. Collage of pasted papers, 44⅞ × 35½". Nationalgalerie, Staatliche Museum, Berlin

FIGURE 58 MAX ERNST. *Loplop Introduces Members of the Surrealist Group.* 1930. Collage and pencil, 19¾ × 13¼". Museum of Modern Art, New York

heritage of the creators of jazz with a statement about the consequences of slavery and racism.

Max Ernst (1891–1978) invented the Dadaist-Surrealist collage in 1918. Sometimes Ernst combined tangible objects in these pieces, but for the most part his collages are a special form of montage.

FIGURE 59 ROMARE BEARDEN. *The Prevalence of Ritual: Baptism.* 1964. Collage of paper and synthetic polymer on composition board, 9 × 11⅞". Hirshhorn Museum and Sculpture Garden, Smithsonian Institution, Washington, D.C.

A connected invention of his was the Surrealist version of *frottage.*

Frottage means "rubbing," which is how Fig. 60 was produced. The artist placed a sheet of paper over variously textured surfaces and rubbed it with a stick of graphite, thereby recording the texture with the paper. Most of us, when we were children, did the same thing with coins, pencils, and writing tablet paper. If sufficient care is taken, it is possible to secure very accurate reproductions of relief forms through this method of drawing. Indeed, a popular and sometimes profitable hobby is doing rubbings of brass plaques (Fig. 61), old gravestones, and other decorative antique surfaces. Anthropologists sometimes utilize the method for recording carvings they would otherwise have to deface or remove.

A characteristic common to many collages, montages, and frottage compositions is the juxtaposition of contrasting kinds of things. Rough-looking textures are put down next to smooth elements while magnified bits of reproduced engravings are framed by photographs and paper doilies. Things rendered

FIGURE 60 MAX ERNST. Frottage from *Histoire Naturelle*, 1926. Private collection.

FIGURE 61 CONSTANCE ADES. *Sir Roger de Trumpington.* 1971. Rubbing of the second-oldest brass in Britain (1289), 95 × 19″. Private collection. The brass itself is in the north chapel of the Parish Church of St. Mary and St. Michael in Trumpington, Cambridge, England.

with different media and printed by diverse kinds of machinery are thrown together in radically antithetical ways. Such adjacency of conflicts tends to point up the peculiar characteristics of individual fragments. Sometimes the distinctions are startling, as when Romare Bearden combines pieces of human beings with shards of statuary. Sometimes similarities receive as much stress as differences; that is the case in Charles Wickler's collage, *Camouflage* (Fig. 62), where the principal variation in the stripes has to do with their degree of regularity. The black and white contrast, as well as the relative scale of white to black, is the same on the background form as upon the zebra's hide.

Designer and critic Roy R. Behrens (b. 1946) has taken the elements of contrast native to collage or montage and adapted them to literary illustration (Fig. 63), but he has deleted properties that for him are distracting or superfluous (such as the contrast between snapshots and diagrams) by reconciling them through high-contrast photography or by drawing on top of the elements he has combined (Fig. 64).

In his book *Art and Camouflage* Behrens analyzed the role of contrast in nature and in art, using the craft of concealment as a clue to the nature of creativity in art. The camoufleur deals in delusions and tries to obliterate our awareness of a "thing" that is being hidden. For a creative act to occur, a "framing" of the act (by actual picture frames, placement in museums, theatrical stages, religious edifices, scientific laboratories, or tones of voice that imply jokes) is essential. Framing "separates acts of creation (which are *labeled* deviant acts) from such phenomena as camouflage, errors, dreaming, pornography, and madness which are unframed deviant acts."[3]

This business of establishing a little universe of order in contrast to something external is not of merely theoretical interest. That designing is deeply involved with the psychology of perception is one of

FIGURE 62 CHARLES WICKLER. *Camouflage.* 1980. Collage and pastel, 30 × 22″. Artist's collection.

FIGURE 63 ROY R. BEHRENS. Design appearing opposite the title page of Jerome Klinkowitz's *The Life of Fiction.* Graphics by Roy R. Behrens. © 1977 by the Board of Trustees of the University of Illinois.

FIGURE 64 ROY R. BEHRENS. Cover illustration for *Art and Camouflage,* The North American Review, University of Northern Iowa, 1981.

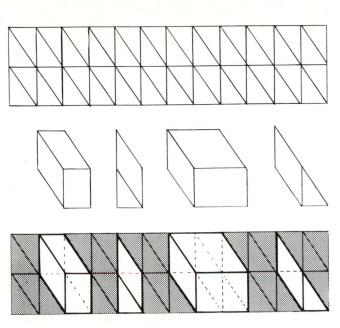

FIGURE 66 JACK OLSON. *Chair.* 1960. Etching, 10 × 8″. Private collection

FIGURE 65 Embedded figures. There are, of course, numerous other possibilities apart from the ones our shading reveals.

FIGURE 67 FRANCOIS ROBERT. Booklet cover. Dunbar Corp.

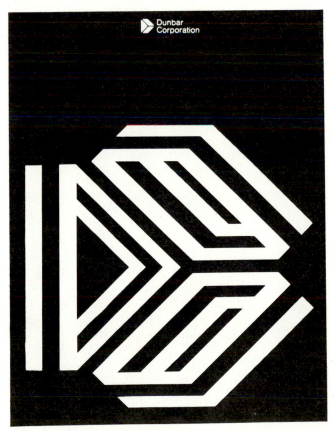

Behrens's main points. He discusses, among other things from the history of psychology, the so-called "embedded figure tasks" studied by Gottschaldt (Fig. 65) in which one is set the problem of discovering, within a complex array of forms, a comparatively simple figure. "When these figures are discerned, a switch of attention would seem to occur, so that what was 'figure' at first (the larger complex whole) must be sloughed in part as 'ground.'"[4] We can, of course, make the task no work at all by shading in what is to be treated as ground. Designers and artists often do just that in order to make clear what might otherwise be very obscure indeed.

The chairback in Olson's etching (Fig. 66) would be scarcely discernible were it not for the decorative treatment he has given the background. A variation in the slant of the stripes blanketing François Robert's brochure cover reveals a new version of the Dunbar Corporation signet, seen in conventional small scale at top center (Fig. 67).

EXERCISE 3-1: Cut-paper Designs

Materials: *Pad of 8 1/2″ × 11″ colored con-*
struction paper
Rubber cement

Tools: *Single-edge razor blade or frisket knife*
Cutting surface

Although Stuart Davis's *Colonial Cubism* looks as though it might have been made from scraps of paper, it is a relatively complex design, all in all. The neophyte would do better to attempt a design along the lines of Fig. 68. Essentially, all of them were fabricated in the following way: Select any sheet of construction paper. Slice it up with the razor blade; just zip a few direct cuts through the paper without attempting to make them "do" anything except move easily and directly. Normally, this will produce much more interesting and authoritative forms than you could deliberately invent. When you have a few pieces, arrange them on another sheet of a contrasting color. Some parts of the pieces may run off the edge and that is all right; you can trim the excess off at will once you have decided upon an arrangement. You don't have to use every shape that you cut out. Quite likely you will have some left over. You might even use only one shape and discard the rest.

For the present, avoid symmetrical dispositions

FIGURE 68 Cut paper designs by students

FIGURE 69 JACK YOUNGERMAN. *Black Yellow Red.* 1964. Oil on canvas, 54 × 42″. Washburn Gallery, New York

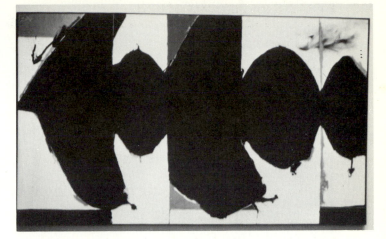

FIGURE 70 ROBERT MOTHERWELL. *Elegy to the Spanish Republic, No. XXXIV.* 1953–54. Albright-Knox Art Gallery, Buffalo, N.Y. Gift of Seymour H. Knox.

of forms, and don't even think about "balancing" the pieces. Instead, as you are studying the arrangements, think about the backgrounds, the negative areas. Strive to give these areas the kind of boldness the positive forms have. You might even want to create a puzzle for the viewer, making it impossible to tell which shapes have been cut out and which result from the whole sheet showing through.

Also, strive for a certain amount of variety in scale. Unity is virtually assured by the facts that you are using only two colors for each design and your cutting gestures will produce forms as much alike as your gestures are. Different people will produce different shapes, just as their gestures will vary. The main risk is that a design will be too monotonous, so vary the size of the scraps and the scale of the negative areas separating them.

When you have produced a number of these designs, expand the color range by using two or three different colors for the cutouts. You will discover that an increase of colors produces an enormous increase in the complexity of the problems you face in arranging the pieces. You can now also make use of shapes for which you had no purpose before.

A number of contemporary artists have done

paintings involving the considerations confronted in this cut-paper exercise. For example, Jack Youngerman (b. 1926) did such a work in *Black Yellow Red* (Fig. 69). Robert Motherwell (b. 1915) too has created impressive works in the same general vein (Fig. 70).

EXERCISE 3–2: Photomontages

> *Materials:* *Same as for Exercise 3–1, plus some picture magazines*
>
> *Tools:* *Same as for Exercise 3–1*

Do precisely as you did above, but instead of slicing up solidly colored construction paper, cut up pictures from the magazines. A little experience with reproductions will doubtless draw your attention to the fact that reproduced photographs tend to exhibit some dominant color overall. That is, color photographs in magazines tend to be greenish, reddish, violescent, beige, or some other color. Take advantage of that tendency. Fig. 71A shows a few student works produced this way.

In Fig. 71B "mosaics" using uniformly square segments of photographs have been reassembled in somewhat disconcerting ways. In these cases all of the segments are portions of the human image, but this device can be used with anything. Try it.

Finally, create a montage of your own, illustrating some prejudice of yours. By "prejudice" we mean a mode of thought in which you tend to judge a thing or a person on grounds that do not necessarily

FIGURE 71A Photomontages by design students

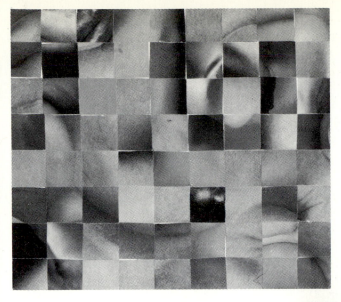

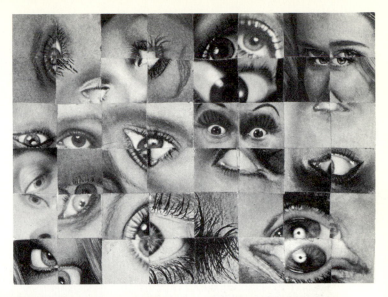

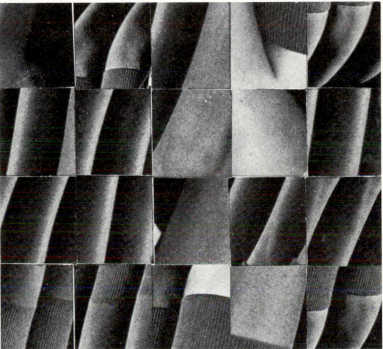

FIGURE 71B Photomontages by design students

have much to do with the individual being assessed. Racism and sexism represent modalities of prejudice. Two other examples: some people imagine that every artist is depraved and there are artists who imagine that all business people are banal clods.

EXERCISE 3–3: A Frottage-Collage

Materials: Newsprint pad, 18″ × 24″ or larger
Tools: Graphite sticks, razor blade, and cutting surface

The result of the combination of textures in Ernst's frottage drawing (Fig. 60) is pictorial but it need not have been. Nonrepresentational patterns can be just as effective. You can also piece together various rubbings and make up a collage of them just as you can from scraps of paper or photographs.

Find as many contrasting textures as you can and then combine them into designs, either by juxtaposing new rubbings of them or by cutting the sheets of newsprint up and doing collages with the fragments.

EXERCISE 3–4: Variety from Sameness

Materials: Same as for Exercises 3–1 and 3–2
Tools: Same as for Exercises 3–1 and 3–2, plus a metal-edged ruler and cutting surface

Fig. 72A is not, as one might at first suppose, a painting. It is a medicine cabinet in one of the authors' bathrooms. A bit more imaginative than the usual mirrored surface, it was inspired by a magazine cover designed by Arnold Saks. In this case a number of quarter circles of the same size have been adjusted with respect to one another in such a way as to make a pattern. From regularity, variety emerged. It's really rather like a grid with some of the units left out. Fig. 72B gives some graphic variations of a similar sort by Vanderbyl.

Devise a shape of some sort and cut out a number of duplicates. Then create an arrangement like the one on the medicine cabinet. Just copying the cabinet design is of no value, though. Use something other than a quarter circle for a motif, perhaps one of the scraps left over from one of the previous exercises.

FIGURE 72A Medicine cabinet

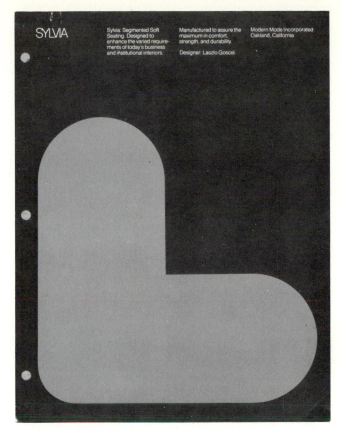

FIGURE 72B MICHAEL VANDERBYL. ModernMode, Inc. Brochure cover, Oakland, California

FIGURE 72C MICHAEL VANDERBYL. ModernMode, Inc. Brochure cover, Oakland, California

Using the same basic shape, create several different arrangements. Use photographic elements as well as solid colored paper elements. Also, do not feel that it is necessary to keep within an 8 1/2″ × 11″ space; if necessary, cut the background sheet to a narrower or fatter proportion.

notes for chapter 3

[1] Piet Mondrian, ''De nieuwe beelding in de schilderkunst,'' *De Stijl,* I. Amsterdam, 1919. English translation by Michel Seuphor in *Piet Mondrian and Work* (New York: Harry N. Abrams, n.d.), pp. 142–43.

[2] The word *montage* also has a special meaning in cinema production, where it refers to a rapid sequence of images used to show the passage of time, mental states, historic associations apart in time or in place, and the like.

[3] Roy R. Behrens, *Art and Camouflage* (Cedar Falls, Iowa: North American Review, 1981), p. 79.

[4] *Ibid.,* p. 68.

chapter 4

Some people would not refer to the Martin (Fig. 16), the Hofmann (Colorplate 3), the Davis (Colorplate 4), or similarly nonrepresentational works as "pictures." They would reserve that word for paintings, drawings, and prints that *depict* something. Thus, the Holbein (Fig. 32), the Degas (Fig. 51), and the Michelangelo (Fig. 53) are "pictures," the Martin, Hofmann, and Davis are "paintings." This usage is, indeed, consistent with lay language, which assumes that pictures "look like something." It also sustains the unquestionable truth that a painting is something painted, a drawing something drawn, and a print a printed image.

A related linguistic convention is sometimes applied to the organization of representational and nonrepresentational works. Probably most artists would grant that what is called "composition" with respect to pictures is a special kind of design. When we speak of the composition of a work like *Colonial Cubism* by Davis, we are in fact talking about the design of the painting. However, when we talk about the organization of a clearly pictorial work such as Holbein's *French Ambassadors,* we are more apt to apply the term "composition" than we are the word "design."

No matter the language, paintings and drawings that provide illusions of reality must be organized—composed, if you will—just as surely as any other. In fact, where organization is lacking, illusion itself is absent. Consider the black spots in Fig. 73A. Rearranged in 73B, they have become letterforms. Similarly, the darks making up the tentacle-coiffured sorceress (Fig. 73C) are meaningless in 73D. The authors selected a comparatively complicated subject for Fig. 73C because the need for

consistency is intensified in elaborate arrays of shapes. Too, while the sorceress is unreal—a fantasy creature compounded of such incongruous elements as a handsome woman and octopus tentacles—the treatment is not unrealistic in its effect. On the contrary, the drawing has qualities not unlike the photograph in Fig. 74A, which looks just like a drawing except for its mechanical objectivity. It was made from the same negative that produced 74B, but it has been processed in such a way as to drop out all intermediary grays. Special techniques can create what look like remarkably accurate pen and ink or crayon drawings (Figs. 74C and 74D). As a matter of fact these *line conversions* have displaced freehand drawing in much advertising art and thereby have vastly reduced the need for illustrators in the field of commercial art.

The comparison of the different versions of the photograph in Fig. 74 should demonstrate the dependency of realism upon bold masses rather than details. Furthermore, it is easy to see how radical simplifications can be applied to relatively more complicated renderings. For example, in Fig. 75 we have begun a sketch of a Mauser machine pistol by reducing it to two values. In effect, though, the light gray stands for what we *might* have reduced to white, and the dark gray is equivalent to what we might have turned into black. Then, actual black and white are superimposed upon the two grays, serving to signify highlights and dark shadow. The looseness of the rendering hints at a rather hasty image of reality. Tighter, more detailed rendering would, of course, elicit a different response. The addition of more values and greater meticulousness could produce an

FIGURE 73 A: Letterforms in disarray; B: Shadow lettering; C: Sorceress; D: Elements of Fig. 73C in disarray.

The principle of organization in both 73B and 73C is the same: the dark areas have been positioned according to the kinds of shadows that given forms would cast were they lighted by the sun's rays slanting down from the upper left. This organization of forms is so effective that it eliminates the need for subsidiary information. We don't have to see the letterforms or the features in the white spaces; we simply fill in the missing elements. Similarly, we believe that the shadows conceal unseen contours and we accept the image as a complete representation.

impression of photographic accuracy. But realistic "impressionism"[1] is itself a special kind of thing. Walter Murch (1907–1967) evoked a kind of poetical sense for plain objects through his rather precise version (Fig. 76) of impressionism; Larry Rivers (b. 1923) has used an even looser kind to conjure up memories of the transient documents of life—currency, cigarette packages, cigar boxes, and similar ephemera (Fig. 77).

If an artist captures the broad, general scheme of a thing, it will be relatively easy to enhance the subject with surface textures and other specifics, but

FIGURE 74 A: High contrast photograph; B: Conventional halftone reproduction of the photograph from which Fig. 74A was made; C: Photographic "line reduction" to produce cross-hatch effect; D: Photographic "line reduction" to produce "mezzotint" effect.

FIGURE 75 A rendering of a Mauser Machine Pistol, first sketched in two values of gray and then elaborated in white and black.

FIGURE 76 WALTER MURCH. Study for *"The Birthday"*. 1963. Pencil, wash, and crayon, 23 × 17½″. Whitney Museum of American Art, New York

no amount of effort can make individual details add up to an authentic-looking image of reality. Such itemization can produce a charming effect, surely. Naive geniuses have built whole *oeuvres* upon it (see Fig. 78), and early Renaissance painters in northern Europe came close to mirroring nature by meticulously itemizing everything they could see.

In his famous painting of the *Madonna of the Chancellor Rolin* (Fig. 79), Jan van Eyck (c. 1390–1441) specifies every detail of the official's robe, his features, the room, and the landscape. Even the Holy apparition of Mary, Jesus, and the little angel is so particularized and detailed as to seem deeply material rather than a spiritual symbol of Rolin's devotion.

The miracles of accuracy attained by Jan van Eyck and the Flemish masters who followed him were of a sort to astonish most viewers, but they also departed from normal vision.[2] Everything in van Eyck looks as if it were being viewed from about four inches away. The Baroque period (c. 1590–1750) produced artists whose realism was supported by a knowledge of scientific perspective and atmospheric effects not available to van Eyck.

Caravaggio (1573–1610) painted his *Entombment of Christ* (Fig. 80) 170 years after Rolin's son became a Bishop and the Chancellor presented van Eyck's painting to the Cathedral of Autun in celebration of his son's attainment. Caravaggio's representation of the world is far more sophisticated than van Eyck's.

FIGURE 77 LARRY RIVERS. *French Money I.* 1961. Oil on canvas, 35¼ × 59″. Courtesy of the artist

FIGURE 78 HENRI ROUSSEAU. *The Dream.* 1910. Oil on canvas, 6′8½″ × 9′9½″, The Museum of Modern Art, New York, Gift of Nelson A. Rockefeller

FIGURE 79 JAN van EYCK. *The Madonna of the Chancellor Rolin.* c. 1433–34. Oil on panel, 26 × 24⅜″, The Louvre, Paris

FIGURE 80 CARAVAGGIO. *Entombment of Christ.* c. 1603. Oil on canvas, 9′10″ × 6′6″, Vatican Gallery, Rome (Alinari-Scala)

It may not be as charming, but it is more knowing when it comes to the facts of visual reality. Nothing in Caravaggio is in such absolute focus as everything is in van Eyck; the Italian master substituted a generalized clarity for the relentless enumeration of minutiae. Proportions in the Caravaggio seem very correct in spite of the low vantage point from which we are viewing the scene.

In Caravaggio's own time many patrons of the Church disliked his persistent portrayal of holy personages as poor commoners—of the sort the Scriptures would lead us to expect—instead of noble patricians like themselves. They called him vulgar. Predictably, later ages defended the artist for his integrity and emphasized his "honest realism."

In 1921 the traditional view of Caravaggio's style was revised. A major scholar, Lionello Venturi, said that he and others had already clarified the events of the artist's life and work but that the "interpretation of his style came first . . . from a great novelist, Aldous Huxley, who in his earliest novel,

Crome Yellow (1921), explains from a cubist point of view the simplified, formalized nature represented by Caravaggio."[3]

Probably it is not easy to see much similarity between a Cubist work (Fig. 81) and the *Entombment of Christ.* Picasso (1881–1973), painter of the Cubist piece, once said that art "has always been art and not nature. And from the point of view of art there are no concrete or abstract forms, but only forms which are more or less convincing lies. . . . Cubism is no different from any other school of painting." The relevance of this statement can be most readily apprehended if we invert the Caravaggio and examine it upside down (Fig. 82).

FIGURE 81 PABLO PICASSO. *Portrait of Ambroise Vollard.* 1909–10. Oil on canvas, 36 × 25½". Pushkin Museum, Moscow.

FIGURE 82 CARAVAGGIO. *Entombment of Christ,* inverted.

When one sees Caravaggio's *Entombment* inverted, one's first impression is of a series of forms looped and knotted together. But the thigh of the dead Jesus appears quite columnar and the lower part of the legs seems to protrude through a rectangular window of shadow. The geometrics of this picture are pronounced once you begin seeking out relationships. Christ's right hand, which appears at the top of the inverted reproduction, is part of a continuous curve that moves through His arm and neck, goes across the shoulders of the mourners who lean over Him, and terminates at the hand of the somewhat theatrical wailer in the background. Her hand, however, initiates another curve which matches the first one. It passes down past the dark rectangle, through Christ's feet to the innermost foot of the apostle supporting His legs. These two "parentheses" bound the figure group. Similar reverberations of forms recur throughout the painting.

FIGURE 83 MICHELANGELO. *The Holy Family* called the *Doni Tondo*. c. 1503. Oil on panel, diameter 47¼", Uffizi Gallery, Florence

FIGURE 84A Sorceress

Almost any apparently realistic painting will exhibit like relationships of forms, but Caravaggio is remarkable for having gone so far without seeming to have *stylized* his figures. That is, the forms seem to conform to nature even though they have, in fact, been modified by his style, which impresses upon natural things an artistic value and geometry.

Michelangelo's *Holy Family* (Fig. 83) is somewhat less naturalistic in its initial impression; his Christchild is infinitely more noble looking than van Eyck's grave infant and his adults are cast in attitudes altogether too statuesque to be human. Still, it is not common practice to call this degree of artifice "stylization." Usually, we reserve the term for the kinds of distortions one sees applied to the human aspects of Fig. 84A as contrasted with the comparatively happenstance spottiness of Fig. 84B. Stylization connotes the imposition upon a natural form of some other, very limited series of forms. The most striking examples of such manipulations of nature are calligraphic renderings in which the flowing line of a quill pen inscribing interlaced circles and spirals is used to suggest fanciful human or animal forms (Fig. 85). Treating the same figures somewhat more three-dimensionally by imposing "carved" facets upon them (Fig. 86) is still within the meaning of the term "stylization." Usually, however, when artists have done something of the sort Picasso did in Fig. 81, they are said to have surpassed "mere stylization" and to have attained *abstraction*. The difference between the two terms is dependent to a large

FIGURE 84B High contrast photograph

extent upon intentions. In Figs. 85 and 86 the artists intended to accomplish nothing more than to represent Cupid, a dog, and stick in ways marked by repetitions of inorganic, decorative shapes. A viewer is somewhat surprised to find these pen lines and plane surfaces used to draw such subjects; it appeals to the sense of "cuteness" most of us possess. (Recall that the word *cute* is a shortening of *acute* and origi-

FIGURE 85 Calligraphic Cupid and dog after Johaan Georg Schwandner in *Calligraphy* (New York: Dover Publications, Inc., 1958) which reproduces designs from Schwandner's 1756 *Calligraphia Latina*

FIGURE 86 Schwandner's Cupid and dog restylized as if carved from wood

The word *stylization* has rather negative overtones, suggesting superficiality, decorativeness, and cuteness, but the kinds of distortions it entails are present in all well-composed depictions of the physical world. The fact is that when the stylization is as subtle as it is in Caravaggio or Michelangelo, we simply accept it as part of formal composition. When it is as patently overt as it is in Cubism and is also as thoroughgoing as it is in Picasso's *Portrait of Vollard,* we call it *abstraction.* Whether you apply the term to something like Grant Wood's style becomes a matter of personal preference.

The word *abstraction* is itself quite ambiguous. It is universally recognized today that *abstract art* is dependent on the idea that artistic value resides in the forms and colors independent of subject content. An abstract work may resemble something else (apples, nude women, a landscape, city streets) but the stress will be on form and color. Some people in the arts feel that the term *abstract art* should be reserved for art work in which some element of reality is preserved—as it is in Picasso's portrait or in Paul Klee's *Landscape with Yellow Birds* (Fig. 34)—and that works like Hofmann's *The Golden Wall* (Colorplate 3) or Rothko's *Orange and Yellow* (Colorplate 10) should be categorized as *nonrepresentational.* Finally, there is the viewpoint of Picasso and the Formalist critics who look upon all art as being primarily abstract, regardless of its superficial appearance.

FIGURE 87 PABLO PICASSO. *Portrait of Ambroise Vollard,* inverted.

If you read about Cubism, you will unquestionably encounter the idea that the Cubist style results from combining separate views of pieces of objects. This notion is sometimes associated with a mystical concept called "the fourth dimension" and *that* conception has been related to Einstein's Theory of Relativity. It's all nonsense. (Einstein *himself* said so, by the way, and both Picasso and Braque denied the multiple viewpoint theory.) Those interested in such philosophical musing can refer to John Adkins Richardson, *Modern Art and Scientific Thought* (Urbana: University of Illinois Press, 1971), pp. 104–27 or Richardson, "Cubism and the Fourth Dimension: A Myth in Modern Criticism," *Diogenes* (Spring 1969), pp. 99–109, or (in a study done quite independently of these) Lynda Dalrymple Henderson, "A New Facet of Cubism: The 'Fourth Dimension' and 'Non-Euclidean' Geometry," *Art Quarterly* (Winter 1971), pp. 411–33.

nally meant to be clever, as when making a "cute" remark.)

Picasso's *Portrait of Vollard* (Fig. 81) does, surely, stylize the physical appearance of the famous art dealer, but it does so in order to attain a more coherent or thoroughgoing organization of the painting overall. Picasso did not stop with fragmentation of an image into V-shaped shards; he carried through a very complicated integration of all of the fractioned parts. (It would actually be more accurate to say that he built the painting from the units than to say that he

broke nature down into bits and pieces.) Picasso sacrifices nearly everything to his desire to make the composition of his paintings their principal subjects; Monsieur Vollard is no more than the pretext for a compositional adventure.

Cubist works done between 1910 and 1912 are very difficult to analyze because they involve such complex spatial relationships. Fortunately, the *Portrait of Vollard* is a bit less complicated than most Cubist paintings of the first rank and is therefore somewhat more accessible. Notice, just below the face of the sitter, projecting out to your right, a dark form set at a slight angle to the picture plane (that is, to the flat surface of the painting). Now, consider the apparent relationship of this form to those adjacent to it. One's first supposition is, usually, that the dark form stands slightly ahead of the other shapes. But does it really seem so? Think again. Perhaps the light area beneath Ambroise Vollard's chin is connected to the dark rectangle, forming with it a prowlike protrusion. On the other hand, the dark shape might conceivably reside behind the lighter one. The ambiguity of the forms is one of the most striking things about the painting, and it is not due to some strange, chaotic transparency; it is a consequence of the ways the individual planes are made to relate to one another.

One important result of such interrelationships is that when you change assumptions about where a form resides in space, you also perceive the rest of the painting in a different way. Let us intensify this effect by inverting the Picasso as we did the Caravaggio (see Fig. 87). A zone of broken panes in the lower right corner of the inversion flows around to the central part of the work, where the light area ties it to another assembly of light gray forms which move up and out of the top right center of the reproduction. The darker sectors on the left and right are related to the light areas in a supportive manner. However, to take a second alternative from the vast multitude Picasso provided for us, consider the two pointed, arrowlike shapes in the upper left of the inverted reproduction. They can be conceived of as part of the series of dark units linked up with that dark plane with which this discussion began. Imagine those darks as advancing forms; think of them as marching over a generally lighter division that passes under them and moves down to fill the lower left portion of Fig. 87. This gives an entirely new character to the forms on the lower right and also summons the "face" of Vollard into a new role.

Not all of these associations may seem plausible to you. Still, they should indicate the degree of seriousness with which Picasso undertook the design of his Cubist portrait. They should reveal, too, why it is that a realistic representation, even a highly styl-

FIGURE 88 CHARLES SHEELER. *City Interior*. 1935. Oil on panel, 22 × 27″, Worcester Art Museum, Worcester, Massachusetts

ized one, cannot attain the same kinds of internal interdependency as an abstract or nonrepresentational work can. This is not to say that the modernist method is the superior one, but only to note that traditional modes cannot achieve all positive things in art.

To do what Picasso did would have been incomprehensible to Caravaggio, whose aim must have been to attain similarly coherent structures without revealing them to any but the most passionately sedulous analyzer of art works. That purpose is very close to what the ex-modernist in Aldous Huxley's *Crome Yellow* hoped for. But the actual, living modernist, Picasso, would have considered the goal of his literary imitation a rather aimless one, inasmuch as Caravaggio had already achieved it.

There have been serious modernists who combined naturalism with abstract perfection of form. One of the most remarkable was Charles Sheeler (1883–1965), an American. As in Caravaggio's work, the adjustments are not immediately evident. However, there is a fundamental difference, deriving from the dissimilar subject matter. Only a specialized engineer is likely to be able even to guess whether the things in *City Interior* (Fig. 88) have been accurately reproduced, whereas even an expert anatomist would be hard put to enumerate the deviations from correctness in *Entombment of Christ*.

By way of contrast, Grant Wood (1892–1942), in *American Gothic* (Fig. 89), reveals the stylization of his world, even as he projects an illusion of truthfulness.

Even photographic realism depends upon the manipulation of certain conventions more than it

FIGURE 89 GRANT WOOD. *American Gothic*. 1930. Oil on beaverboard, 29 3/8 × 24 7/8″, The Art Institute of Chicago. Friends of American Art Collection

The trident pitchfork is repeated in the stitching of the bib overalls, in the symmetry of the man's face, in the "Carpenter Gothic" window upstairs, and in other places throughout the work. Similarly the daughter's apron is topped with a rickrack decoration, the arc of which matches the fork's bend and also repeats the contours of her face, the trees, and the top of her father's head.

(Father and daughter? That's right. Like most people, the authors had supposed that this picture was of a farmer and his wife, largely because we didn't really look at these people modeled on Wood's sister and his dentist. But in 1980 Mrs. Nan Wood Graham, who was thirty when she posed for the picture, said that her brother had intended to depict a farmer and his daughter, not a farmer and his wife as is commonly believed.)

does upon the direct imitation of reality. For example, in Edward Weston's photograph of weather-beaten wood, painted white (Fig. 90), the various shades, ranging from the black of the doorknob to the brightness of clear sunlight, represent varying degrees of light and shade. Obviously. But to know this is to know how to make *sense* of the photograph. We understand the conventions of photographic imagery and are therefore capable of "reading" the variegated pattern as a record of an old church door at midafternoon on a sunny day.

The apparent illusionism of any realistic-looking picture depends upon our willingness and ability to pretend it resembles what it doesn't, upon our acceptance of the rules or conventions of a certain kind of image. A photograph won't look like reality to you unless you know what the various patches of dark and light are supposed to stand for. People from cultures in which the representational conventions of the West are unknown cannot make sense of photographs; for such a person the grays are just arbitrary blotches on a piece of paper. Even our seemingly common-sensical assumption that the top of the picture represents "up" is a convention. This is *not* the same as saying that the naturalism of photography is completely arbitrary, that all images are equally subjective in nature, for people from societies without a tradition of realism in art can be taught to understand what photographs mean very quickly when they are instructed to trace outlines with their fingers. That is

because the silhouettes of human figures, animals, trees, and so on do retain a constant relationship between the directly seen and their two-dimensional representation. A horse and rider viewed against a setting sun presents the same silhouette to the human eye and to the camera lens. On the other hand, no photograph of a setting sun, no matter how superbly printed, is going to make you squint into the sun's rays.

We cannot teach all of the conventions of representationalism in a book any more than a team of grammarians could instruct readers in all of the conventions of the English language. We can, however, convey a few basic ideas.

Since most of you come from cultures sufficiently imbued with European imagery to take the fundamental conventions for granted, we shall not belabor them here. Nonetheless, it is important to understand that pictures are never realistic, actually, and that organization is essential if clear communication is to take place between the artist and the viewer. Weston's photograph is not merely a record of a segment of California in 1940. It is also a work of art, one that has been quite beautifully composed. If we superimpose a grid on it, certain correspondences will come into focus (see Fig. 91). *This is NOT because Weston employed such a grid pattern to compose the picture.* It's simply the case that grids that divide works into thirds frequently have a general coincidence with artistic subdivisions (see Fig. 92).

FIGURE 91 EDWARD WESTON's photograph divided into thirds

FIGURE 92 Some masterpieces divided into thirds: A: MICHELANGELO. *The Holy Family;* B: VAN EYCK. *Madonna of the Chancellor Rolin;*

Here is a selection of masterpieces upon which we have superimposed a triadic (3-part)grid; one can scarcely help being impressed with how often major components of the works coincide with the divisions and crossings. Even the Picassos fit this grid rather well. Most surprising, pehaps, is the Mondrian, since nothing at all corresponds to one of the grid lines. This is a pretty definitive demonstration that Mondrian, in effect, devised new compositional possibilities. On the other hand, many of his proportions subscribe to the ancient formula for beauty, Euclid's "Golden Section."[4]

a

b

c

d

e

C: CARAVAGGIO. *Entombment of Christ*; D: PICASSO. *Portrait of Ambroise Vollard*; E: PICASSO. *Girl Before a Mirror*; F: MONDRIAN. *Composition in Black, White and Red*;

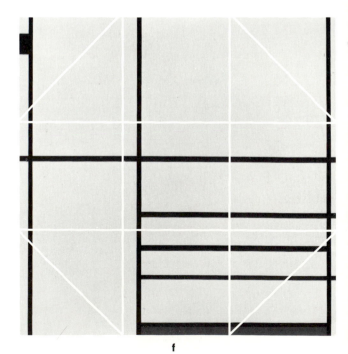

f

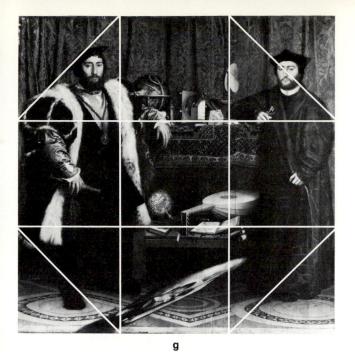

G: HOLBEIN. *The French Ambassadors;* H: DEGAS. *At the Milliner's.*

Many artists, illustrators, and designers begin their sketches and preparatory studies by dividing the space to be used into thirds. It is not a rule that one should do this. It isn't even a good idea always to do it. However, it *is* a handy way to avoid some of the more dubious arrangements.

Usually, it is best to begin a design with uneven relationships, such as thirds and fifths, rather than with even ones, like halves and quarters. However, even a symmetrical paradigm based on the quartering of the picture space will prove highly adaptable. Fig. 93 takes the same works as Fig. 92, but now we

FIGURE 93 Some masterpieces divided into quarters and subdivided by diagonals: A: MICHELANGELO. *The Holy Family;* B: VAN EYCK. *Madonna of the Chancellor Rolin;*

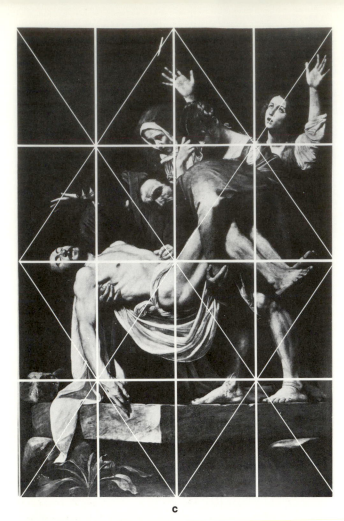

c

d

e

C: CARAVAGGIO. *Entombment of Christ;* D: PICASSO. *Portrait of Ambroise Vollard;* E: PICASSO. *Girl before a Mirror;* F: MONDRIAN. *Composition in Black, White and Red;*

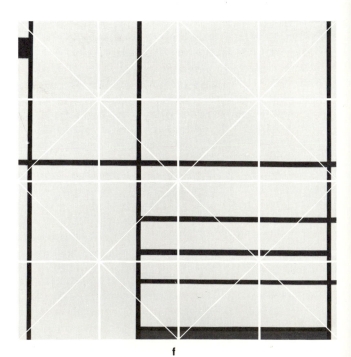

f

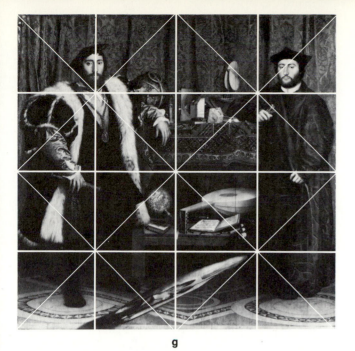

g

h

G: HOLBEIN. *The French Ambassadors;* H: DEGAS. *At the Milliner's.*

have applied a system of evenly spaced horizontals and lateral zones split by diagonals. As you can see, it doesn't have much bearing on the Mondrian. However, with respect to some of the other works—especially the Holbein, Caravaggio, and *Portrait of Vollard*—the correlations are quite startling.[5]

It should not surprise you to observe that what works out for the great masters also works out for popular illustrators, cartoonists, and graphic designers. Figs. 94 and 95 show that it does in fact.

It is clear, too, that symmetrical compositions need not result from a symmetrical scheme. Using a one-third division of space to establish the axis of a dominant form (as in Fig. 96) leads to asymmetrical arrangements. Contrariwise, if one employs the triangular cells of the quartering system as containers for figures and objects (as in Fig. 97), the effect is somewhat informal even though the process itself is highly formalized.

Selectivity, too, is a major factor in the creation of a coherent picture. In Fig. 98 we have simply *cropped* the larger view in different ways. To crop a picture, particularly a photograph, is standard practice in the graphic arts. Most newsphotos are treated this way and it is practically inevitable that designers doing magazine layouts and advertising design will find it necessary to eliminate portions of pictures, either to improve the composition, to stress a part of the larger whole, or to fit the picture into a rectangle of a different shape than the original print.

Associated with the matter of selectivity is that of dominance or emphasis. Many pictures have a dominant area which years ago used to be called the "center of interest." When we first began working with the people in Fig. 98, we actually sought to make them the dominant elements of the picture. It is by no means necessary always to have so dominant a focus for the viewer's attention. In many paintings one wanders throughout the work, paying little attention to any particular portion. That is true, for example, of the nonrepresentational works we have dealt with thus far. It is also true of numerous landscapes and still lifes done in conservative styles. Usually, in a representational work, though, there will be a center of interest. In Michelangelo's *Holy Family* (Fig. 99) the emphasis is generally the zone containing Mary's face and the Christchild. Jean-Léon Gérôme (1824–1904), in his painting *The Gray Eminence* (Fig. 100), drives our attention to the little Capuchin monk, Father Joseph (1577–1638), who, as the confidant and agent of Cardinal Richelieu (1585–1642), was believed to be the second most powerful man in France during the indifferent reign of Louis XIII (1601–1643).

The painting is a mature version of French Academicism and fits the paradigm of a quartered subdivision very well, as our lines in the diagram show. Such compositional armatures are a hallmark of *Academicism,* a mode of teaching art which left as little as possible to the individual imagination and

a

c

FIGURE 94 Examples of popular art divided into thirds: A: N.C. WYETH. Illustration for *Treasure Island* by Robert Louis Stevenson (New York; Charles Scribner's Sons, 1911, reissued 1981); B: DAVID JOHNSON. Illustration for "My Father, Myself," by Sheila A. Warner in *Ms.* (May 1981); C: JOHN ADKINS RICHARDSON. Cover of *Papers on Language and Literature* illustrating an article on T.S. Eliot's influence on Hemingway, Fall 1978; D: MILTON GLASER. Illustration for "The Politics of Talking in Couples" by Barbara Ehrenreich in *Ms.* (May, 1981).

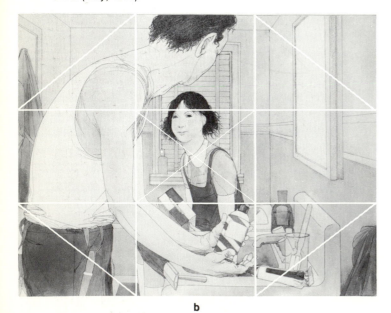

b

d

a

Papers on Language & Literature

c

FIGURE 95 Examples of popular art divided into quarters and subdivided by diagonals: A: N.C. WYETH. Illustration for *Treasure Island* by Robert Louis Stevenson (New York: Charles Scribner's Sons, 1911, reissued 1981); B: DAVID JOHNSON. Illustration from "My Father, Myself," by Sheila A. Warner in *Ms.* (May 1981); C: JOHN ADKINS RICHARDSON. Cover of Papers on *Language and Literature* illustrating an article on T.S. Eliot's influence on Hemingway, Fall 1978; D: MILTON GLASER. Illustration for "The Politics of Talking in Couples" by Barbara Ehrenreich in *Ms.* (May, 1981).

b

d

FIGURE 96 Preliminary layouts based on subdivision of space by thirds

FIGURE 97 Preliminary layouts based on subdivision into quarters further subdivided by diagonals

a

b

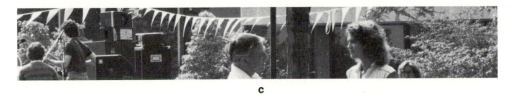

c

d

e

FIGURE 98 A: Panoramic photograph; B: Full length close-up of figures in 98A; C: Radical close-up of figures in 98A; D: Well-composed full length detail of figures from 98A; E: Medium long shot of figures from 98A

If we wished to represent two people conversing on a campus, we probably would not represent the scene as a panorama (Fig. 98A) because, while the setting is nice, the figures are hardly more than adjunct elements of their surroundings. A closer view of the figures is called for. Figure 98B, though, is a particularly unimaginative and uninteresting way to present them. It's the way Uncle Fred takes photographs of Aunt Sadie when they are on vacation; the figure is not very intriguing, and not enough of the background appears to provide an appealing context. Usually, a radical close-up yields more dramatic appeal than a full-length view. When you can preserve a hint of the setting, which the background gives us in Fig. 98C, the effect is rather good. Figure 98D is much better composed than the other full-length shot or the close-up, but it doesn't lose the figures. If we wished to convey something about the scale of the campus relative to the people while retaining some of the individuality of the people involved, a medium-long shot, with the figures off to one side, might be the way to go (see Fig. 98E).

made the most of devices, like formal subdivision, which can be learned by rote and can be taught to nearly anyone.

An earlier, less humorous Gérôme, *Police Verso* (Fig. 101), reveals the same sureness of purpose and grandeur of setting as *The Gray Eminence*.[6]

Still today, illustrators for the mass media make use of many of the devices Gérôme and other nineteenth-century conservatives used (see Fig. 104). On the other hand, fashion dictates that modes other than the standard pseudo-realism of old-style history painting come into prominence. Every new movement in the art world has had some influence on commercial illustration. So have nostalgia fads which revive fashions from the past in some new form. At the moment of writing, a certain kind of watercolor technique, associated with lithographic color separations during the earlier part of this century, has

FIGURE 99 MICHELANGELO. *The Holy Family* called the *Doni Tondo*. c. 1503. Oil on panel, diameter 47 1/4″. Uffizi Gallery, Florence

FIGURE 100A
JEAN LEON GEROME. *The Gray Eminence (L'Eminence Grise)*. 1874. Oil on canvas, 38 3/4 × 25 3/4″, Museum of Fine Arts, Boston (Bequest of Susan Cornelia Warren)

FIGURE 100B
Gerome's *The Gray Eminence* with the dominant movements diagrammed.

Gérôme has set a perfect stage for his little drama. The glorious costumes of the obsequious courtiers and the general magnificence of the staircase contrast startlingly with the studious, distant cleric who seems so utterly incorruptible in his plainness. Gérôme's painting technique is absolutely stunning, the color marvelous, and the drawing superb, in a picture that is just over a yard wide. The painter has focused our attention on the monk by turning him into the detached point of what is, in effect, a triangular sweep of figures on the staircase. Fig. 100B diagrams movements essential to the effect. The fact that Father Joseph wears a smock that is uniformly dull and dark also helps emphasize his silhouette against the heavy tapestry on the landing. The direction of everyone's glance causes ours too to speed across the composition to the far right edge of the picture.

FIGURE 101A
JEAN LEON GEROME. *Police Verso.* c. 1859. Oil on canvas, 38 1/8 × 58 5/8″. Phoenix Art Museum, Phoenix, Arizona

FIGURE 101B Gerome's *Police Verso* with dominant movements diagrammed

In this painting a rather paunchy, middle-aged, but obviously skilled warrior looks to the Colosseum audience for a judgment. The Vestal Virgins, seated on Caesar's left, turn "thumbs down" along with the rest of the crowd, thus sealing the vanquished combatant's doom. Gérôme has arranged the elements of the arena to focus our attention on the group of athletes and the Virgins. As in *The Gray Eminence,* many elements converge upon the center of interest, in this case the victor. His pose, combined with the beseeching gesture of his fallen opponent, completes a visual circuit tying the Vestals and the crowd beyond them to the gladiators (see Fig. 101B). Both of the main groups are areas of high contrast where light forms are juxtaposed with dark shapes—the carpeted hanging in front of the Vestals' box, the shaded armor of the contestants, and the satin banner ahead of Caesar that is background for the principal gladiator.

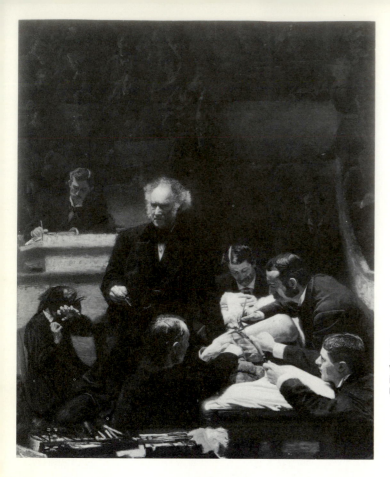

become the rage—so much so that it seems about to become passé on its own account.

A few recent illustrations appear in Fig. 105. If you would like to see something more up-to-date than any book can possibly show, walk through a large record store; some of the most imaginative work being done in the field of graphic illustration today is for record jackets. Tastes in popular art change constantly, and by the time this book is edited, typeset, proofed, printed, bound, and distributed, *whatever* we use to exemplify the currently successful will have become dated.

FIGURE 102 THOMAS EAKINS. *The Gross Clinic.* 1875. Oil on canvas, 96 1/4 × 78 5/8″. Jefferson Medical College, Thomas Jefferson University, Philadelphia

FIGURE 103 FERNAND LEGER. *Three Women (Le Grand Dejeuner).* 1921. Oil on canvas, 72 1/4 × 99″. The Museum of Modern Art, New York. Mrs. Simon Guggenheim Fund

a

b

FIGURE 104 Conservative contemporary illustrations
A: N.C. WYETH. "I said goodbye to Mother..." Illustration for *Treasure Island* by Robert Louis Stevenson (New York: Charles Scribner's Sons, 1911, reissued 1981.) Reprinted with permission of Charles Scribner's Sons © illustration 1911, 1939.
B: Cover illustration for *A Valley Called Disappointment* by Robert Bell (New York: Ballantine Books, 1982).
C: DAVID JOHNSON. Illustration for "My Father, Myself" by Sheila Weller in *Ms.* (May, 1981).

c

FIGURE 105 Modernist contemporary illustrations
A: Cover illustration for *The Sea Wolf* by Jack London (New York; Signet Classics, 1964).
B: KATHERYN HOLT. Illustration for "1955, or, You Can't Keep a Good Woman Down" by Alice Walker in *Ms.* (March, 1981).
C: MILTON GLASER. Illustration for "The Politics of Talking in Couples" by Barbara Ehrenreich in *Ms.* (May, 1981).

b

a

c

EXERCISE 4-1: Transferring Printed Imagery with Minimum Equipment

Materials: *Lighter fluid*
White drawing paper
Picture magazines printed on slick paper
Tools: *Medium hard drawing pencil (H, 2H)*

Artist Robert Rauschenberg (b. 1925) has for a long while integrated hazy ''memories'' of magazine reproductions into his paintings and prints (Fig. 106). They sometimes look almost as if he had been able to copy them by hand in some quick, magically impressionistic way because scribble marks track the places the image appears. It is the result of a clever trick which can be used by anyone who wishes to create striking contrasts.

Simply take a page from a magazine printed on so-called ''slick paper,'' such as *Life, Time, Playboy, Ms., Cosmopolitan, Newsweek, Parade,* or *Ebony.* Select a picture and, in a well-ventilated space safe from sparks or flame, squirt lighter fluid onto the reproduction in the magazine. When it is thoroughly moist but not saturated, place the page face up on a flat surface and lay a sheet of drawing paper on top of it. Then draw lines back and forth on the paper, moving quickly yet firmly and covering the entire area over the fluid-dampened picture. When you peel the drawing paper away from the magazine page, you will observe a rather pale mirror image of the reproduction. The volatile lighter fluid has loosened the greasy particles of ink in which the picture had been printed and the pressure of the pencil pressed them into the fiber of the relatively absorbent drawing paper.

Transfer images of this kind lend themselves to combination, either by direct overlapping and juxtapositioning or by the collage method (Fig. 107). Clearly, they can also be combined, as Rauschenberg has done, with photographs, drawings, painting, and lettering.

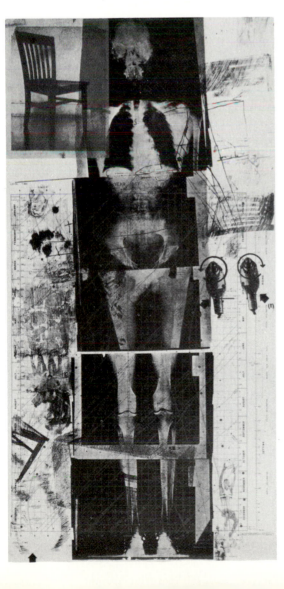

FIGURE 106 ROBERT RAUSCHENBERG. *Booster.* 1967. Color lithograph and silkscreen, 6′ × 3′. Courtesy Leo Castelli Gallery, New York.

FIGURE 107 Lighter fluid transfer exercise by a student of design

FIGURE 108A HANS HOLBEIN. *The French Ambassadors.* 1533. Oil on panel, 81¼ × 82¼". (Detail) National Gallery, London

FIGURE 108B After a detail from Hans Holbein's *The French Ambassadors*

EXERCISE 4-2: Reductive Conversion of Darks and Lights

Materials: *White drawing paper*
Fine arts reproduction of an old master painting
Tools: *Pen and brush*
India ink

Take a reproduction of a relatively complicated work of art such as van Eyck's *Madonna of the Chancellor Rolin* and reduce it to harsh contrasts of black and white as we have done with the fragment from Holbein's *French Ambassadors* (see Figs. 108A and 108B). Select a fairly large reproduction from which to work and try to retain as much of the character of the image as possible. This may prove to be a good deal trickier than it looks, and no two people will find themselves in complete agreement about where white should leave off and black begin. Be certain of this at least: you can't copy, or even trace the image off casually, filling in areas with ink and just leaving it at that.

The best procedure is first to shade the forms in with one value of penciled gray, then estimate whether a given shadow should be larger, smaller, contoured differently, and so forth. Once you have decided on what the best simplification will be, ink the gray areas in in solid black. Try to be as closely observant as possible; it is easy to overlook a lot in another's work.

EXERCISE 4-3: Cropping for Effect

Materials: *Picture magazines or photographs*
Tools: *Red fiber-tip pen or grease pencil*

Take a number of photographs or pictures reproduced in magazines and crop them to fit within rectangles of the following sizes:

1. 5″ × 8″
2. 1″ × 7″
3. 3″ × 9 1/2″

It will *not* be necessary to cut the pictures up. Cropping is done by using *crop marks* on the margins of the photograph or on the board on which the picture or photograph is mounted. Fig. 109A shows how crop marks work; whatever falls outside of the rectangle formed by connecting all of the marks is to be deleted. This technique enables the technicians to see what is to be included without having to destroy the photograph itself. (Thus, we were able to use a single print of Fig. 98 to produce all of the other versions of the image from 98B through 98E.)

If your reproductions are too small or too large to work within the dimensions we have designated—as we intended them to be—you can still crop them for those dimensions by *scaling* the pictures to the required spaces.

The most convenient way to do this is by the *diagonal-line method* diagrammed in Fig. 109B. With

FIGURE 109A Crop marks

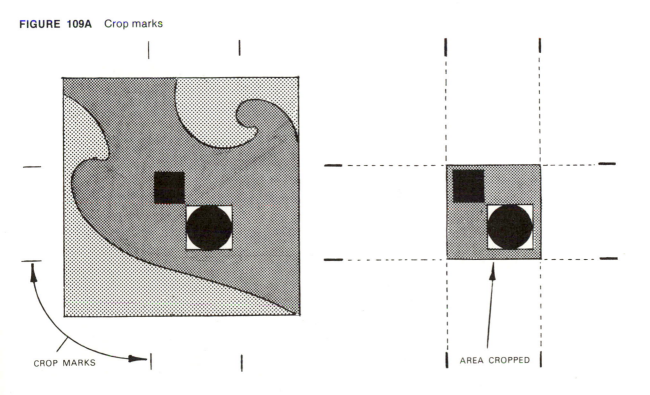

CROP MARKS

AREA CROPPED

pencil, mark off the limits of the rectangle you need to fill. Draw a diagonal line from the bottom left through the upper right corner. Any rectangle that has its top and right side parallel to the one drawn and that has the same diagonal through its corners will be *exactly* the same shape as the initial one. *All rec-*

tangles based on a common diagonal have the same proportions no matter how large or small they may be. Irregular forms can be scaled in the same way by enclosing them in a rectangle which touches all of the outermost parts of the silhouette, as the bear has been treated in Fig. 109B.

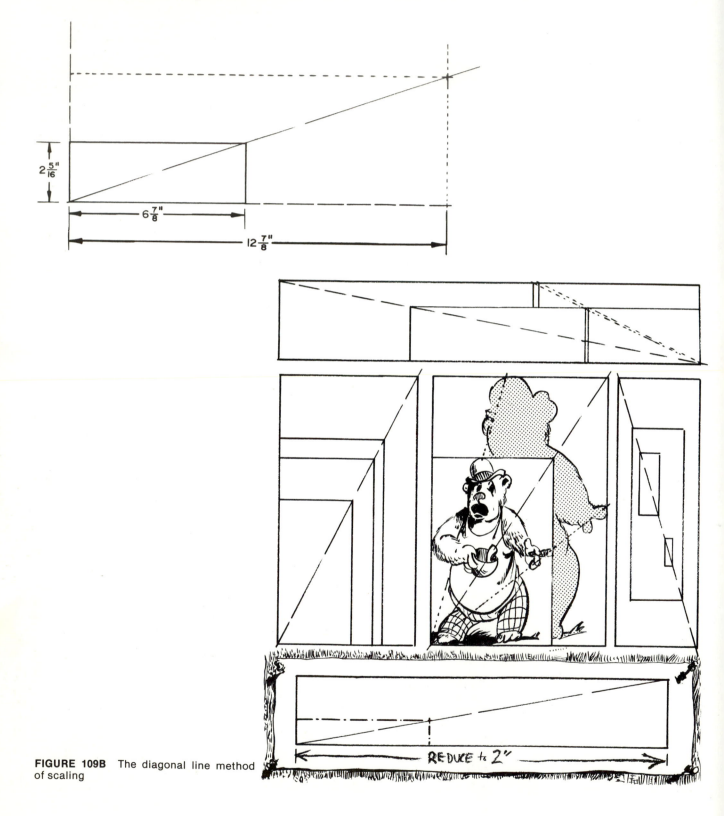

FIGURE 109B The diagonal line method of scaling

EXERCISE 4-4: An Illustration

Materials: *White drawing paper*
Tools: *T-square*
Ruler
Triangles
HB drawing pencil
Pen and ink
Felt pens
Fiber-tip pens
Other drawing or
painting media

Compose a picture containing figures, still life objects, and/or landscape elements in which the distribution of elements has been planned according to an underlying subdivision of the space into thirds or quarters. Employ the following procedure:

1. Obvious, but necessary: Decide on a subject.
2. Do at least ten ''thumbnail sketches'' (Fig. 110) in which you try out different arrangements in rectangles no more than 3 inches in the largest dimension. Detail is of no consequence but make sure that the rectangles have pretty much the same shape and that the masses you draw into them are proportioned according to the space they will take up in the final drawing.

3. In composing the picture avoid the primitive space arrangement exemplified in Fig. 111A, where whatever is more distant is higher up on the paper and whatever is nearer to the viewer is lower down. There are times, certainly, when you might wish to use such a disposition of elements to imply, say, innocence or maplike clarity of position, but generally it should be avoided just because it *is* naive. It is an essentially inexpressive approach to the depiction of the world, a simple intellectual construction in which a mapping-out of positions is substituted for visualization of things. Become accustomed to thinking in terms of overlapping and diminution (see Fig. 111B) instead of in terms of the amateur's ''map space.''
4. Select from your batch of thumbnail sketches one you prefer and make a more elaborate sketch or layout of at least 6 inches in the smallest dimension. Enlarge the thumbnail sketch by expanding the cells just as you did in creating the earlier grids. This drawing should be about as finished and detailed as Fig. 112.
5. Do a final rendering of the illustration in the medium of your choice.

FIGURE 110 Thumbnail sketches

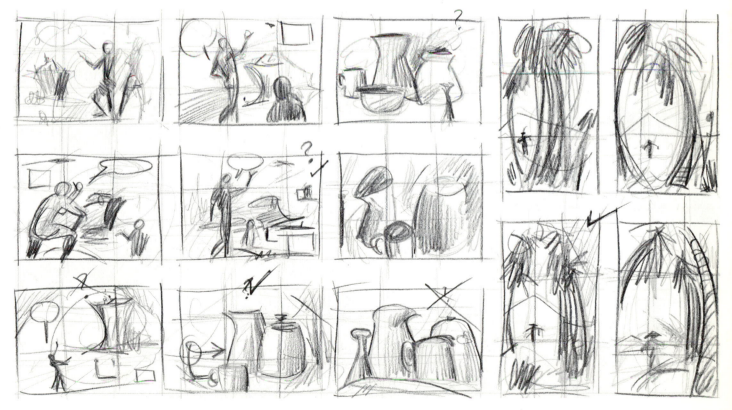

FIGURE 111A Primitive "mapping" spatial arrangement in which nearer things are lower in the picture and more distant things higher.

FIGURE 111B Spatial arrangement based on overlapping and diminution of things.

FIGURE 112 Preliminary layout

notes for chapter 4

[1] The term *impressionism* is, like many designations in common currency among artists, applicable to different things. When capitalized, *Impressionism* refers to French Impressionism, a movement in painting that originated during the last third of the nineteenth century. Attempting to attain an ultimate kind of Realism through an exact analysis of color, pictures in this style are painted with short, brightly colored dabs of paint. The subject matter tends to be concerned with landscape, cafe scenes, theatre and ballet, the casual life of artists, and still life. Uncapitalized, as we have used it in this part of the text, *im-*

pressionism means imagery which captures a true-to-life effect in a quick, rather sketchy manner. In art circles the word almost never is used to mean what many laypeople believe it means—a subjective distortion of reality. That is usually some form of *expressionism,* according to current usage.

[2] Jan van Eyck attained such verisimilitude in details partly because of the medium he used. It is *oil paint.* He and his brother Hubert have long been credited with the invention of oil paints. Recently, it has become fashionable to deny their claim, but this argument is more semantic than real, for the van Eycks were the first *masters* that we know of able to produce paints that ran the full range from utter opacity to complete transparency and could be thick, dense, or highly fluid. Furthermore, there is no evidence to suggest that the brothers lifted the idea from some more obscure painter. For centuries egg yolk had been the standard vehicle used to bind powdered pigment particles to one another and to the surfaces being decorated. The resulting paste, known as *tempera paint,* presented some problems: it is sticky, it dries almost at once, and it can never be completely opaque. Painters had used oil to make the egg less sticky and slow down drying. What the van Eycks did was replace the egg entirely and use oil in its place.

[3] Lionello Venturi, *Four Steps Towards Modern Art* (New York: Columbia University Press, 1962), pp. 24–25.

[4] The "Golden Section" is a proportion such that the longer part is to the smaller as the whole is to the longer. A "Golden Rectangle" is a rectangle in which those sections form the sides and ends. There is something peculiarly natural and appealing about the proportion; it occurs all through nature—even in shells and microscopic cell structures—and is frequently encountered in art.

[5] Some theorists of art, calling their theory "Dynamic Symmetry," have attempted to join mathematical subdivision to abstract proportions based on such ratios as the Golden Section in order to devise mathematical systems of compositional analysis. The subdivisions included in this book are older, Academic schemes which make no pretense to being anything more than encouragements to neophytes and crutches for the experienced.

[6] Gérôme is one of those supremely gifted artists whose reputation suffers from his popular appeal and from the vast changes in artistic fashion that have succeeded in eliminating anecdotal painting from practically everything except paperback books and novelty posters. Curiously, Gérôme himself has been immensely influential in two ways. The first has to do with his personality. Thomas Eakins (1844–1916), the highly regarded American realist (Fig. 102), imitated Gérôme's teaching when he returned from Paris to the Philadelphia Academy. Among Gérôme's other students was the modernist Fernand Léger (Fig. 103), who praised him for not interfering with students' personal styles. Also it was Gérôme who recognized the positive value of Henri Rousseau's naiveté (see Fig. 78) and discouraged him from studying at the Academy, telling him to guard the originality of his innocence.

Gérôme's second influence is upon popular art, particularly the art of the cinema. *The Gray Eminence, Police Verso,* and similar historical works established in the minds of succeeding generations an image which was easily adapted to story illustration and movie epics. The scale relationships, melodramatic contrasts, and elaborate blocking of figure groups for visual effect were taken over by commercial illustrators and motion picture directors, who could treat the same subjects on a far grander scale than could have been managed on a theatre stage.

part 3

elements

line and form

The things common to all visual art, what are called the ''elements'' of art, are usually identified as *line, form, light, color, space,* and *texture.* Already, in some fashion, we have touched upon all of the elements in the previous chapters, for they are so basic to design that without them neither imagery nor pattern could exist. To label them is merely a convenience; the attributes described by the words merge into one another so that it is impossible to draw hard and fast distinctions among them. They are like such interrelated components of music as melody, harmony, rhythm, meter, and timbre. Few of us think of rhythm as being separate from melody when we hear a favorite song. We rarely imagine what the melody would be like given a different timbre. Musicians may, of course, but even they do not really think of the elements as being separate entities, although they use the words to make it easier to talk about certain aspects of composition and performance. Similarly,

artists and designers use a word like ''form'' to direct attention to certain features of art works. To a layperson they often seem to be playing fast and loose with the English tongue, as when they talk about ''line'' in a work like Mark Rothko's *Orange and Yellow* (Colorplate 10). But a common-sense approach can help us understand what is meant.

Consider Fig. 113. It demonstrates that line is to form somewhat as weight is to mass, faith is to belief, or weather is to climate. These pairs are not identical, but they are always found together. Lines can be narrow stripes, like *A, B,* and *C,* or can be outlines of shapes like *J, I,* and *H.* In fact, what is described as ''line'' might be any number of things (see Figs. 114 and 115). Of course, alignment involves line, too (see Fig. 116). Line in this sense is frequently termed ''movement'' by artists. It can be an effective way of tying images together in a coherent fashion. A certain internal harmony and grace can be

FIGURE 113   Lines and Forms

Which of these would you refer to as ''lines,'' in casual conversation? Which would you call ''forms''? Most of us would call ''lines'' the marks that are so thin that you don't really pay attention to their ends. These have a noticeable length—they terminate, after all—but the ends are not an important feature of their appearance. The ones labeled *H, I,* and *J* have tops and bottoms more than mere ''ends.'' With respect to *A, B,* and *C* our eye seems to run up and down even though *A* is obviously narrower than *C;* with respect to *H, I,* and *J* it seems to run *around* them. We might go so far as to say the latter have lines around them, that their edges are the same as lines. This is quite true. A line can describe the edge of a form or it can be a long thin mark.

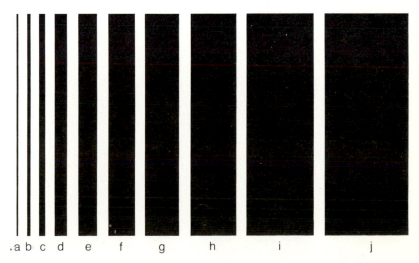

.a b c d e   f   g   h   i   j

FIGURE 114   The arrows point out three kinds of boundary lines

A line can be an outline in the sense of a boundary marking off an area. It might be, in that case, an edge or a fencelike mark enclosing or designating a zone. Skylines, treelines, hairlines, and any other sort of outlines involve the same thing—line.

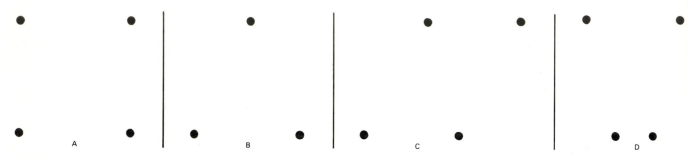

FIGURE 115   Invisible lines

In this diagram we see what may be taken for "invisible lines" since the dots suggest geometric forms described by connected lines. Here we have (A) a square, (B) a triangle, (C) a parallelogram, (D) and a trapezoid.

FIGURE 116A   Alignments

FIGURE 116B   Linear analysis of Michelangelo's *Holy Family* called the Doni Tondo (see Fig. 99)

Here we have a line of dots and then some lines of spheres. This kind of relationship is easily assimilated to the kind of linearity seen in Michelangelo's *Holy Family*—what artists call "line movements." The alignments of converging forms tend to direct our attention to the upper portion of the painting and to keep it within the ambit of the disc.

a

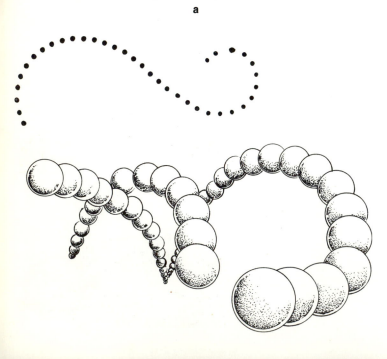

b

FIGURE 117A RAPHAEL. Detail of *The Expulsion of Attila.* 1513–14. Fresco. Stanza d'Elidoro, Vatican, Rome (Alinari-Scala)

FIGURE 117B Figure 117A with diagram of continuities.

ensured by establishing continuous movements. A great master of this sort of continuity was the Renaissance master Raphael (1483–1520). His *Expulsion of Attila* (Fig. 117) in the Vatican is a figure composition of great integrity. In an associated way, we use the term "line" to refer to the general flow of forms. In Fig. 118 all of the automobiles have the same outline—that is, they have identical silhou-

FIGURE 118 The silhouettes of all of these cars are the same—they are derived from the Porsche 924—but the "lines" of the Exoticar, the FIA/4 racing car, and the Clydemobile are not the same as the outline all of them share.

FIGURE 119A MARIE CASSATT.

Mother's Kiss. 1891. Drypoint etching in color, 13⅝ × 8¹⁵/₁₆″. National Gallery, Washington, D.C.

FIGURE 119B Cassatt's print outlines the directions edges would take were the figures treated in terms of masses instead of in terms of thin, dark forms. Notice, too, the very obvious continuities in the child and supporting hand and through the woman's dress and child's thigh.

ettes—but the lines of each are different.

This kind of usage is typical in art and design; when we speak of line we are usually talking about movements. Most often such movements are along the edges of forms. That is, they are lines of boundary. Sometimes, though, they are as diffuse and generalized as the movement around the hazy rectangles in the Rothko. The meaning is fairly obvious when one speaks of lines existing in something like Mary Cassatt's drypoint etching of a mother and child (Fig. 119).

FIGURE 120 The drypoint of the mother and child, redrawn with a heavy, regular line, a varied brush line, a sketchy set of connected marks, a combination of delicate pen marks and a heavy boundary, and a scratchboard imitation of woodcut produces, in each case, a quite different effect.

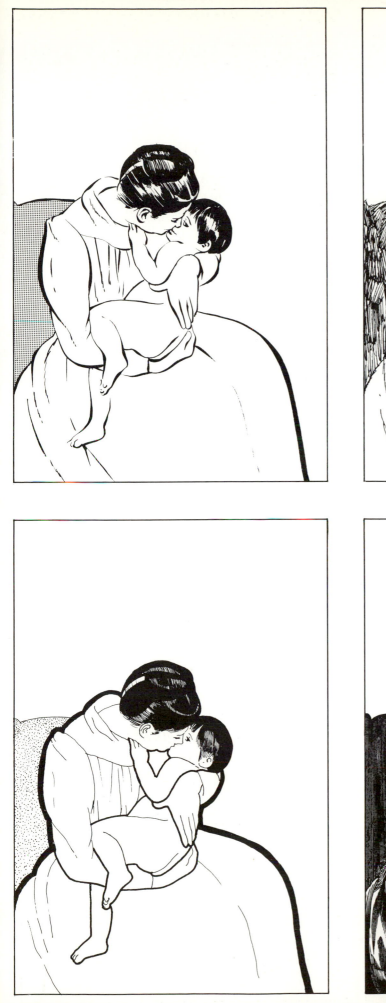

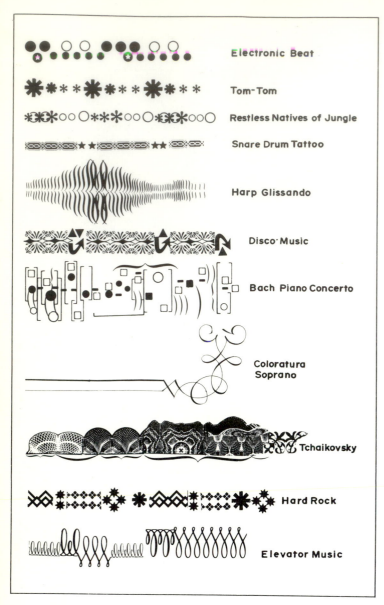

FIGURE 121 RI. *A Little Night Music for Saul.* 1982. ARRESSICO.

Synesthesia is the technical term for a situation like this in which one kind of stimulus (here visual) produces a sensation associated with another mode of perception (here auditory). A taste might have a smell, of course. A feeling may have a color or a given melody a texture. Poets, artists, and composers have made extensive use of such relationships. So, of course, have advertising artists and other designers. The small format advertisement by Hal Tench evokes seasoned richness of fine food through a heavily stylized representation of a steaming pot.

Drawn lines, that is, thin forms corresponding to edges and to other movements, have their own peculiar characteristics, depending upon the implements used to make them and the way a given artist has put them down onto a surface. Cassatt's figures, redrawn with different kinds of marks, produce quite distinctive impressions (see Fig. 120). Even very simple, standardized line references have

lyrical potential in the hands of a trained artist. Ri's *A Little Night Music for Saul* (Fig. 121) is a vivacious example of such a combination. Also one can nearly smell the aroma of a bubbling pot in Fig. 122.

Clearly, the different modes of rendering the human figure in Figs. 123, 124, and 125 are associated with different feelings about what is being seen. Picasso's drawing is spare and sure-handed even though it exaggerates differences of scale, whereas Ingres's line seems directed by a precision so exact as to leave no room for deviation. Ben Shahn's portrait of nuclear physicist J. Robert Oppenheimer is still less exact than Picasso's but it, too, seems to be very accurate in its evocation of a person whose fineness of judgment is held captive by complicated, tragic circumstances of history. Shahn's Oppenheimer is deformed by constraints and the lines from which he is built are snagged and broken.

In one sense, the status of the drawn lines in the three portraits is progressively more independent of the subject. That is, the Ingres drawing is obviously a drawing of a person and not the man himself; yet, the line is very close in appearance to the kind of outline

FIGURE 122 HAL TENCH, Designer. DICK ATHEY, Art Director. Kingsmill Golf Club Advertisement. Webb & Athey, Inc. 1980.

FIGURE 123 JEAN DOMINIQUE INGRES. *Portrait of Docteur Robin.* Pencil, 11 × 8³/₄". Courtesy of the Art Institute of Chicago. Gift of Emily Crane Chadbourne

Three artists working with drawn lines produce very different effects, depending upon the accuracy, exaggeration, or distortion introduced into the proportioning of the figures and variations in the character of the marks they have used.

FIGURE 125 BEN SHAHN. *Dr. J. Robert Oppenheimer.* 1954. Brush and ink, 19¹/₂ × 12¹/₄", The Museum of Modern Art, New York

FIGURE 124 PABLO PICASSO. *Ricciotto Canudo Montrouse.* 1918. Pencil, 14 × 10³/₈". Collection The Museum of Modern Art, New York. Acquired through the Lillie P. Bliss Bequest.

FIGURE 126   PABLO PICASSO. *l'Aficionado.* 1912. Oil on canvas. Kunstmuseum, Basle.

FIGURE 128   AGNES MARTIN. *The Tree.* 1964. Oil and pencil on canvas. Museum of Modern Art, New York

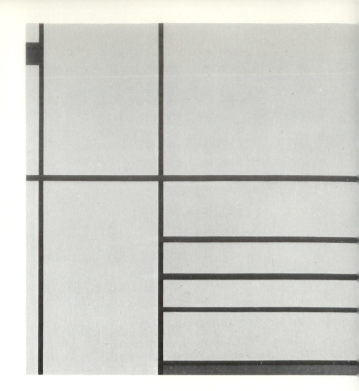

FIGURE 127   PIET MONDRIAN. *Composition in Black, White and Red.* 1936. Oil on canvas, 40¼ × 41″. The Museum of Modern Art, New York. Gift of the Advisory Committee

one would achieve by doing a sensitive tracing of Doctor Robin's image onto a sheet of glass. Picasso departed from such a pattern; his marks could not be matched up with any view of the subject from any viewpoint. Shahn's marks are still more divorced from the real appearance of Oppenheimer than Picasso's are from his sitter's semblance. In fact, Shahn's lines are really a kind of autonomous or self-sufficient design which happens to bear a vague resemblance to a real person. Picasso's cubist paintings (Fig. 126) were an earlier yet further step toward autonomy. Some works of art are entirely bereft of any connection with objects in the real world. Mondrian's *Composition in Black, White, and Red* (Fig. 127) is, in effect, nothing more than a structure of long thin marks on a white surface. So are Martin's grids (Fig. 128) and Pollock's drip paintings (Fig. 129). These works are, of course, *nonrepresentational.* That is, they do not have any relationship to material objects in the mundane world before the fact. You may be able to relate the stripes in a Mondrian to metal bars in architecture (Fig. 130) by Mies van der Rohe (1886–1969), but that is largely because they have a common aesthetic behind them, not because Mondrian was trying to model his paintings after buildings.

Sam Francis (b. 1923) did brush paintings (Fig. 131) containing effects inspired by Pollock. They featured paint laid on with a brush so loaded that the pigment trickled down from the individual strokes, producing a striking contrast between the fat brush stroke and the thin drips streaming down (Fig. 132). During the following decade Roy Lichtenstein (b. 1923) rang changes on this device by painting pictures in which comic strip mannerisms were used to imitate, by painstakingly finicky means, what Pollock and Francis had achieved spontaneously (Fig. 133). Although derivative of something seen, Lichtenstein's "strokes" are about art. Artists have not been slow to surpass Lichtenstein's ploy, however; Mike Smith (b. 1936) has used calligraphic, flat strokes of a brush to produce a three-dimensional effect that is at once illusionistic and spontaneous (Fig. 134).

Three-dimensional art works may have the same independence from ordinary material objects. Constantine Brancusi (1876–1957) called his sculptures by titles that evoke specific referents. *Bird in Space* (Fig. 135) is an exquisitely crafted representation of the theme. It does not really resemble a bird at all, but it anticipates the sleek aerodynamic forms invented many years after the sculpture was cast.

Because the Brancusi is so internally compact, the different aspects of it that we might observe as we walk around it are not strikingly dissimilar. That is, we can be fairly confident that our reaction to a single viewpoint will hold for all other viewpoints as well.

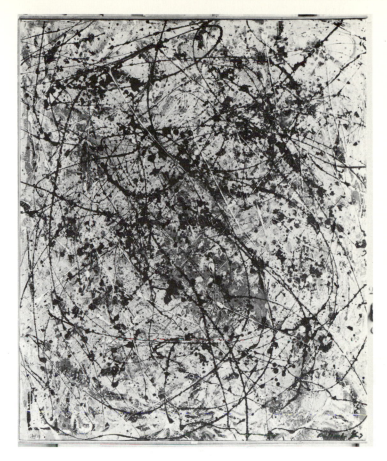

FIGURE 129
JACKSON POLLOCK. *Reflection of the Big Dipper.* 1946. Oil on canvas, 43⅝ × 36¼". Stedelijk Museum, Amsterdam

FIGURE 130 MIES VAN DER ROHE. Crown Hall, Illinois Institute of Technology, Chicago. 1952–56.

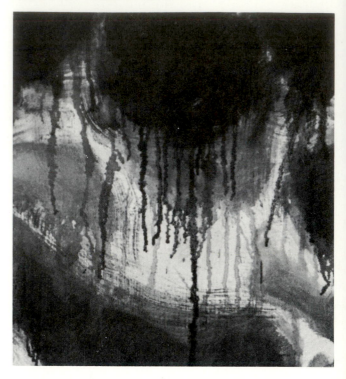

FIGURE 131 SAM FRANCIS. *Red and Black*. 1954. Oil on canvas, 76⅞ × 38⅛″. The Solomon R. Guggenheim Museum, New York

FIGURE 132 Detail of Figure 131

Art is here used to comment upon art. Sam Francis made brush strokes which took advantage of the tendency of a brush pressed hard against a canvas to squeeze out fluid which then trickled down the surface. In effect, he created a new kind of line/form effect by using sloppiness in a controlled, inventive fashion. Roy Lichtenstein, in a satirical comment on this kind of thing, drew pictures of carefree strokes and splatterings with the inherently finicky methods of the technical illustrator. Michael Smith, experimenting with calligraphic use of western materials, gives us a genuinely spontaneous stroke that leaps into space and seems to burst out of the picture plane in the way Lichtenstein's cartoonlike rendering only hints at.

FIGURE 133 ROY LICHTENSTEIN. *Brushstrokes*. 1965. Oil, 68 × 80″. Whitney Museum of American Art, New York

FIGURE 134 MICHAEL J. SMITH. *Traveler "K"*. 1978. Oil on paper, private collection.

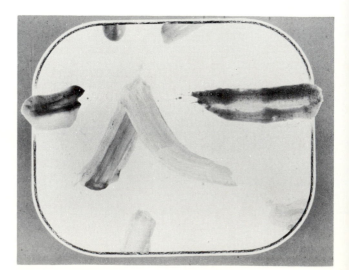

FIGURE 135 CONSTANTINE BRANCUSI. *Bird in Space.* 1924. Polished bronze, height 49¾". The Philadelphia Museum of Art. Louise and Walter Arensburg Collection

That is not the case for other types of sculpture, however. Mark di Suvero's *XV* (Fig. 136) reveals an astonishingly varied character depending upon where one stands relative to it. The simple geometry of the work gives each view a coherence and interrelationship with other views although the results are quite startling at times.

Attainment of such coherence in sculpture that represents a subject that is itself dramatic is less often encountered. Indeed, the ability to establish a fluid relationship of line and form in three-dimensional work such that each view of all of the endless possible views is in harmony with the others is a major problem for the traditional sculptor. To solve such problems in a work of monumental scale, the artist works out solutions in sketches of different sides of the piece and perfects them in the small-scale model, called a *maquette* or *bozzetto,* which is used to establish the proportions of the full-scale statue. Art historian Joy Kenseth has provided us with an analysis of the way one great sculptor, Gian Lorenzo Bernini (1598–1680), accommodated his works in the Borghese Palace in Rome to the relationship between "the fiction he has made and the beholder of that fiction,"[1] that is, to the way in which the positioning and conception of a given piece evokes a series of views which correspond to the story it illustrates. Her views of Bernini's *David* (Fig. 137), as it would have normally been seen where originally located, reveal David's expression becoming "increasingly taut and fierce and his body more compact."[2] The effect of her remarkable series of photographs is astonishingly similar to animation as we encounter it in the cinema. As we move from the figure's left to his right side, our journey "leads us to a point where we become both physically and psychologically aligned with the *David.* Like the biblical hero, we turn our heads to sight Goliath and like David, too, we become potential champions against the Philistine."[3]

The full effect of Bernini's *David,* or of Suvero's *XV* cannot be secured from photographs reproduced in a book made up of flat pages, no matter how in-

FIGURE 136 MARK DI SUVERO. *XV.* 1971. Steel, 21′7″ × 26′11″ × 23′11″. Laumeier International Sculpture Park, St. Louis. On loan, Courtesy of Oil & Steel Gallery, New York

This sculpture is very simple in principle; it consists of industrial I-beams forming an *X,* a *V,* and an *I.* Yet, as the viewer circumnavigates the piece, the associations among the simple, if massive, components take on an unpredictably dramatic impact.

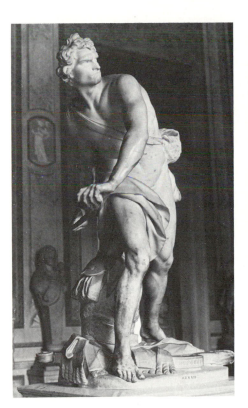

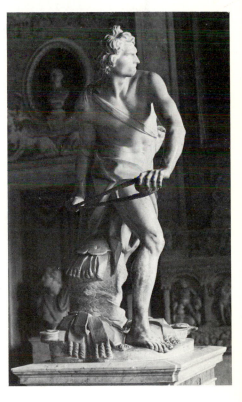

FIGURE 137 GIAN LORENZO BERNINI. *David.* 1623. Marble, height 67″. Galleria Borghese, Rome. (Photos: Joy Kenseth)

Notice the way in which the lines and forms flow into one another as the successive aspects of this sculpture are revealed. For instance, David's left arm begins as a curve parallel to the right upper thigh. By the last view it has begun to fall into a similar relationship with the inner side of the right leg and the line separating the abdomen from the left leg. In the same way, Bernini brought other forms into continuous harmony with one another. Virtually all first-rate representational sculpture reveals the same kind of coherence although it is rare to find it unveiled as dramatically as it is in the *David.*

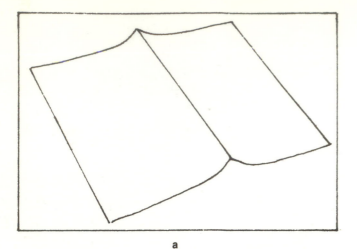

a

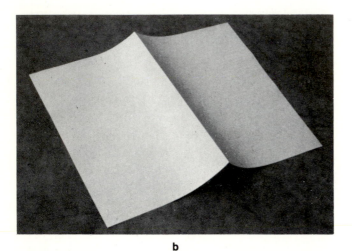

b

c

FIGURE 138 A: A line drawing of the creased sheet of paper in Fig. 138B; B: A photograph of a sheet of paper. The crease through the center is revealed to us by the contrast of dark and light and the contour of the edge; C: A pencil rendering of Fig. 138B. This drawing carries over into the representation essentially the same dark-light relationships contained in the photograph.

geniously the separate views are displayed. The only way one knows that something illustrated in this book is three-dimensional instead of two-dimensional is through the light and shade revealed in the photographs. Indeed, the only thing that makes it possible for you to see the book at all or read words upon the pages is the contrast of darker spots with lighter ones. Contrast between dark and light makes vision possible; when all is black, one is blind, and when everything is of the same brightness, one cannot see. Fig. 138 compares a line drawing (Fig. 138A) with a photograph of the model (Fig. 138B) and a pencil rendering of the same thing (Fig. 138C). If you take a sheet of paper and fold it through the center, then let it rest on a gray, flat surface in rather dim sunlight, it will present essentially the same appearance as images in B and C.

As we noted previously, in connection with Edward Weston's photograph (Fig. 90), no photograph is truly realistic. However, convincingly real-looking pictures, such as the photograph, do preserve some of the principal relationships that obtain for vision. Thus, the silhouette in reality is the same as the one caught on film or traced by the artist's hand. The organizing principle for the shadow lettering (Fig. 139) is derived from the natural world, where sunlight falling in parallel rays lights the sides of things in a consistent fashion and also throws shade along the same pathways away from the light source. In Fig. 140A whatever is to our left is lighted and whatever is to the right is in shadow. These shadows are not utterly black because air itself is reflective and holds light even in the shade. (If the scene were on the moon, where there is no air to speak of, the shadows *would* be black.) Artificial light, emanated by torches, lanterns, electric bulbs, and the like, radiates so that the things in Fig. 140B are lighted on the side toward the flame. This difference between artificial and natural light is obvious to everyone although most people have never really thought about it.

The Italians have a word for light and shade in art. They call it *chiaroscuro* and it has come to have some specialized uses apart from its general meaning (see Fig. 141). Again, however, although chiaroscuro derives from our experience of light and shadow in

FIGURE 139 Shadow lettering

SHADOW

FIGURE 140A   Natural lighting

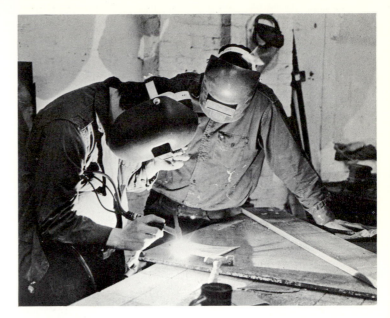

FIGURE 140B   Artificial lighting

FIGURE 141A   The nomenclature of chiaroscuro

(A) The nomenclature of chiaroscuro. The *highlight* represents a reflection of the light source; it is almost pure white. (On a dull surface the highlight will not be reflected clearly but will become a part of the *light* area.) The *penumbra* is the transition from light to what is known as the *umbra,* sometimes known as the "core of shadow." The *cast shadow* is the shade of the object on another surface and is usually the darkest of the components. The final element is *reflected light,* which is a secondary light, usually resulting from light reflected from a background surface.

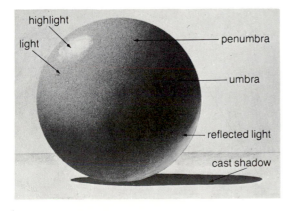

FIGURE 141B   MICHELANGELO. *The Holy Family* called the *Doni Tondo.* c. 1503. Oil on panel, diameter 47 1/4 ". Uffizi Gallery, Florence

(B) Here is a good example of chiaroscuro handled in a skillful, traditional manner. The reflected light is used to bring the Madonna's left upper arm out from the background. Michelangelo has even introduced an unrealistic nimbus of light on Joseph's knee behind Mary's back, bringing her forward in contrast to her husband; that bright light glowing on his knee is utterly implausible but it is very effective. Cheating on nature in this way is quite commonplace in art.

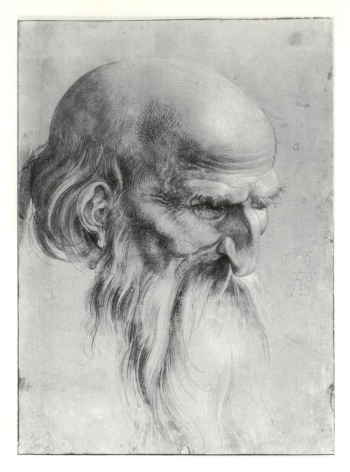

FIGURE 141C ALBRECHT DURER. *Head of a Disciple.* 1508. Brush, ink, and wash, Graphische Sammlung Albertina, Vienna.

(C) The kind of drawing we see in Fig. 141A is properly described as "being in chiaroscuro" since all of the elements of light and shade are provided. However, it is not what is referred to in art circles as a *chiaroscuro drawing* because it was drawn on white paper. Dürer's wonderful rendition of a man's head, on the other hand, is an example of chiaroscuro drawing in the strict sense. The artists worked on tinted paper—usually gray but in Dürer's case blue—indicating shadows with a dark crayon or ink and noting the lightest places with white crayon or paint.

the natural world, illusion does not depend upon subtleties of shading (sometimes called "modeling") details so much as upon fundamental patterns of dark and light such as those in the shadow lettering and in high-contrast photographs (Fig. 142). Too, some ways of using light and shade in drawing do not correspond to the way things look at all, but they are nonetheless quite effective. These techniques depend upon contrast, intellectual consistency, and diagrammatic thoroughness. In Fig. 143 some objects are shown as they might appear under ordinary sunlit conditions. Fig. 144 shows the same things rendered according to the premise that light surfaces are advancing ones and dark surfaces receding. Fig. 145A involves the quite arbitrary alternation of graduated panels of dark and light. In fact most artists resort to

FIGURE 141D HENRICH GOLTZIUS. *A Sea Goddess.* c. 1588–89. Chiaroscuro woodcut. Museum of Fine Arts, Boston

(D) There is also a woodcut technique called *chiaroscuro woodcut* in which light in the print is indicated by white paper, semishade by some medium tone, and deep shadow by black or extremely dense gray. The effect of such woodcuts has been imitated in other graphic media and in painting and drawing as well.

FIGURE 142 High contrast photograph

The extreme contrasts of black and white, representing light and shade, are fundamental to the illusions of chiaroscuro. Big relationships are primary. If the general effect is plain, the addition of highlights, penumbra, and reflections is comparatively easy. But no amount of fussy detail will add up to an illusionistic effect if the large-scale patterns are incorrect.

FIGURE 143

Objects seen under normal conditions, rendered according to the principles of chiaroscuro.

subtle versions of these approaches at least some of the time. The barn in Fig. 145B does not look entirely implausible although careful analysis reveals that it departs from natural lighting (Fig. 144) quite considerably in favor of the formula of Fig. 145A.

What can be done with continuously graduated values of dark and light can, of course, be achieved with various size marks put down in varying densities per square inch on white paper (see Fig. 146). *Halftone* reproduction (Fig. 146A), which reduces grays to tiny dots of black and white for printing in books like this one, does exactly that. So does the artist's handmade version of halftone, *stippling* (Fig. 146B). *Cross-*

FIGURE 144

The same objects shaded according to the principle that lighter areas of a given surface stand for protrusions and dark areas represent receding edges or indentations. The lighter the area, the nearer it is to the viewer; the darker, the farther away. This effect is quite powerful and convincing although it is not consistent with any real light source, not even one directly in front. (Frontlighting would not, for example, illuminate the inside of the bowl in this way.)

FIGURE 145

In Fig. 145A we have the same objects as in Fig. 144, but this time they have been submitted to a simple alternation of graduated light/dark contrasts. This looks rather bizarre in the diagram. However, as you can see from the *wash drawing* of the old barn (Fig. 145B), varying the values so that not every corner is equal in intensity with every other corner makes an otherwise quite unrealistic treatment seem relatively plausible. We have already noted that Michelangelo made some use of arbitrary lighting in the Doni Tondo (see caption 141B).

a

b

a

b

c

d

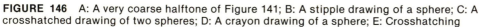

e

FIGURE 146 A: A very coarse halftone of Figure 141; B: A stipple drawing of a sphere; C: A crosshatched drawing of two spheres; D: A crayon drawing of a sphere; E: Crosshatching

In doing this kind of rendering, a rather systematic procedure is usually best. Notice the two versions of the sphere in Fig. 146C; one of these is done with ruled lines, the other with short, freehand strokes. In both cases the illustrator built up the darks by adding lines at different angles from the ones that were put down first. Fig. 146E diagrams this process. Distributing the dots of stippling and laying down the gray zones for a crayon drawing involve essentially the same kind of procedure.

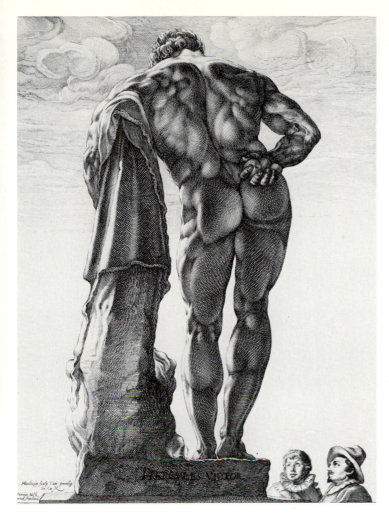

FIGURE 147A HEINRICH GOLTZIUS. *The Farnese Hercules.* c. 1592–93. Engraving. Museum of Fine Arts, Boston

FIGURE 147B Detail of Fig. 147A

The engraved line which results from a sharp groove carved into metal by a v-shaped tool has the capacity to fatten and taper off to a mark so thin as to be invisible to the unaided eye. In this work the artist has followed the forms of an ancient statue which had been discovered in 1540. It is a Roman copy of an original done about 350 B.C. by the Greek sculptor Lysippus. It was restored and set up in the courtyard of the Farnese Palace in Rome. When Goltzius's engraving of it is seen as expected, from a slight distance, the skein of lines models the powerful musculature of the mighty hero, Hercules, whom Lysippus had represented exhausted by the eleventh of the twelve labors eventually assigned him. (He is holding the golden apples of the Hesperides, for which he has not only killed the dragon Ladon but also held up the heavens for Atlas.) If one steps more closely, to study the skin surface of the first and foremost of the Greek superheroes, one is confronted by a linear maze of dazzling complexity.

hatching (Fig. 146C) is merely the straight-line version of the same thing. Crayon drawings (Fig. 146D) substitute relative densities of granular particles for individual marks. (Of course, crayons and pencils may be used to cross-hatch as well as to lay in zones of continuously graduated grays.

Certain printmaking techniques—notably copperplate *engraving,* in which the printed line results from ink resting in a groove cut into the surface of a copper plate by a sharp steel tool called a burin—combine linear characteristics of contour, thickness, and density to convey a very complete sense of shape and chiaroscuro as well as an intricacy that is entrancing on its own account. Heinrich Goltzius's *The Farnese Hercules* (Fig. 147) is a superb example. A highly contrasting impression, but one in which the special qualities of the printed image contribute a refreshing variation on line and form, is *woodblock* or *woodcut* (Fig. 148A), in which the lines that are cut

away with chisels and gouges become the negative, white areas and the surface left alone produces the black lines in the printed image (Fig. 148B). Emile Nolde (1867–1956) so carved a plank of wood that when inked and pressed against a sheet of rice paper, his *Prophet* (Fig. 149A) appeared. The white cuts in the lower left (Fig. 149B) have no particular character; they are simply long gouges. However, the black forms which result in the space between the white ones have a very striking quality. It is a quality that can be imitated with *scratchboard,* a cardboard coated with a uniform thickness of white clay. The artist draws upon this surface with a brush and India ink, then scratches away parts using needles, knives, and special tools designed to cut parallel grooves. Fig. 150 is a scratchboard rendering done in imitation of *wood engraving.* (Wood engraving is exactly like woodcut except that the latter makes use of the plank grain of the wood while the former is carved into the end grain, permitting much tighter control and greater detail.)

Fig. 151 features a number of prints and drawings which employ the different approaches to line and form herein surveyed. In Fig. 152 are relevant illustrations from various publications.

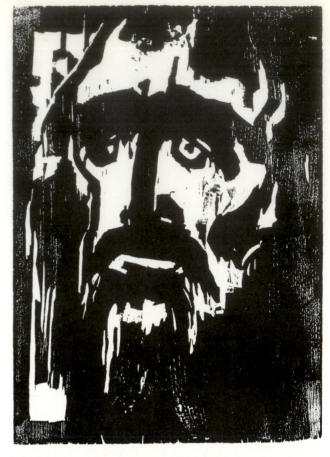

FIGURE 149A EMILE NOLDE. *The Prophet.* 1912. Woodcut, 12¾ × 8⅞″. National Gallery of Art, Washington, D.C. Rosenwald Collection

FIGURE 148A ALBRECHT DURER. *The Four Horsemen of the Apocalypse.* c. 1598. Woodcut, approximately 15¼ × 11″. Metropolitan Museum of Art, New York. Gift of Junius S. Morgan, 1919

FIGURE 148B Detail of Fig. 148A

FIGURE 149B Detail of Fig. 149A

Nolde's woodcut is very rough and direct compared with traditional versions of the medium such as Dürer's. However, like the fifteenth-century picture, the one from the early twentieth century has also a special kind of line resulting from the fact that in the woodblocks the lines are the negative, uncarved surfaces and the white spaces the positive zones. That is, the artist produced the work by doing exactly the opposite of what he at first appears to have done.

FIGURE 150 JOHN ADKINS RICHARDSON. Scratchboard illustration for *Sou'wester* Magazine. 1979.

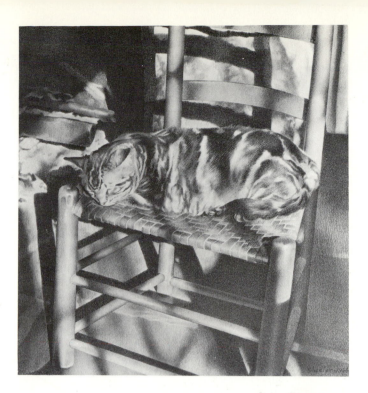

FIGURE 151A CHARLES SHEELER. *Feline Felicity.* 1934. Conte crayon, 22 × 18″. The Fogg Art Museum, Harvard University, Cambridge, Massachusetts. Louise E. Bettens Fund.

FIGURE 151B JOHN ADKINS RICHARDSON. *Self-Portrait.* 1981. Pen and ink, 15 × 9″. Private collection

FIGURE 151C AUGUSTE RODIN. *St. John the Baptist.* Pen and ink, 12³/₄ × 8³/₄″. The Fogg Art Museum, Harvard University, Cambridge, Massachusetts. Grenville L. Winthrop Bequest

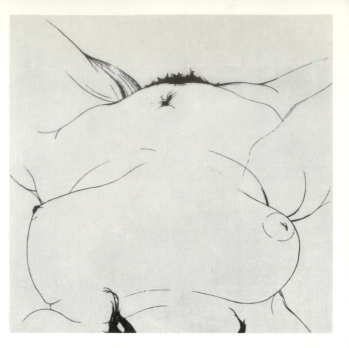

FIGURE 151D CHARLES WHITE. *Preacher.* 1952. Ink on card-board, 21⅜ × 29⅜". Whitney Museum of American Art, New York

FIGURE 151E HENRY KNICKMEYER. *Reclining Nude.* 1965. Engraving, private collection

FIGURE 151F ERNEST BARNES. *Untitled.* 1961. Woodcut, private collection

FIGURE 151G ROBERT MALONE. *Self-Portrait with Aquarium.* 1974. Etching, 24 × 36". Private collection

FIGURE 152A
OSWALDO MIRANDA. Illustration for a humorous story on how an obese woman outwitted three lovers, from the newspaper *Diario do Parana*, 1981.

FIGURE 152B PAUL SIEMSEN. Pablo Picasso self-promotion poster for a studio-printer. The portrait has been done in four different weights of type. © The Word/Form Corporation, 1978

FIGURE 152D
BILL SANDERSON. Scratchboard illustration. Pentagram Design

FIGURE 152C
R.J. SHAY. *Colonel Custer.* Notice that the artist has substituted skulls and crossbones for the brass buttons of the Union Cavalry jacket.

FIGURE 152E JOHN ADKINS RICHARDSON. Scratchboard drawing of D.H. Lawrence. Southern Illinois University Press

FIGURE 152F R.J. SHAY. Pen and ink drawing of Groucho Marx. 1980.

FIGURE 152G JOHN MILLIGAN DESIGN, INC. Catalogue cover for *Prime Computer*

FIGURE 152H ROBERT STEARNS, Art Director. STAN BROD, Designer. Exhibition Catalogue, Contemporary Arts Center, Cincinnati, Ohio

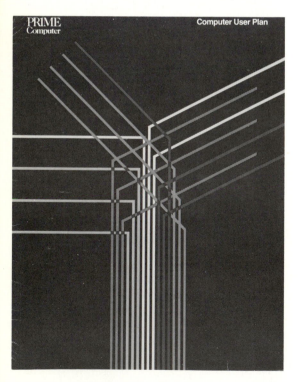

EXERCISE 5-1: Dot to Dot

Materials: *Newsprint sheets 18″ × 24″*
India ink

Tools: *Drawing stick, various brushes, pencils, graphite sticks*
Yardstick

With a pencil, make between twelve and twenty dots on a sheet of paper, spotting them about at random but distributing them over the page. (That is, don't cluster the dots in one place and leave large sections of the paper blank.) Then connect the dots, using the yardstick and a 4B drawing pencil. Join each dot to at least one other dot or to as many as seems necessary to produce an interesting design. The lines may cross one another.

Repeat this exercise, but with the following variations:

1. Use the graphite stick (2B) and draw the connecting marks freehand rather than with a yardstick.
2. Use broken lines in any medium to connect the dots.
3. Do freehand brush lines using various sizes and types of brushes.

EXERCISE 5-2: Inventing Marks

Materials: *White drawing paper*
India ink

Tools: *Pens, brushes, twigs, strings, sticks, twists of paper, leaves, etc.*

Fig. 153 is an assortment of marks made with different kinds of implements, some conventional artist's tools such as pens and brushes, some "just things" that will make some sort of mark. The snail-like curls at the top were done with a technical fountain pen and *French curves*. The circle below them was made with pencils stroked out over a disc of tracing paper. The ribbon running down from the middle of the circle was started with two pens held together. Another experiment with pens held together begins at the bottom center of the page. The strange "squiggly" pair of lines going out to the right of the ribbon was made with a wooden matchstick dipped in ink and wiggled as it was pulled across the paper. Other marks were made with strings dragged along, with torn paper stencils and fiber-tip pens, by squirting a blob of India ink onto the paper and tipping the paper around so that it crawls over the surface. The heavy free strokes on bottom right were done with a Q-tip dipped in ink. The rather more elegant marks just

FIGURE 153 Marks made with various kinds of implements

A

B

C

D

FIGURE 154 Forms suggesting (A) elegance (B) power (C) boredom (D) speed

above were done with a matchstick and a number 4 red sable brush which has the ability to make a line that fattens and tapers. Scruffy old brushes and twigs produced other effects. White lines were made by a pencil going over marks made by a dead ballpoint pen.

Make up a sheet of marks of this kind for yourself, using everything you can think of, including printed material from magazines, catalogues, and so forth.

EXERCISE 5-3: Feeling in Forms

Tools and Materials: *Same as above*

Lines and marks can indicate moods, sensations, and even personalities. Ri's *A Little Night Music for Saul* (Fig. 121) looks the way things sound. Try to sort out the marks done for Exercise 5–2 according to some of the adjectives below. That is, match a given mark to a specific term:

feeble	soft	sinister
brittle	elegant	menacing
powerful	flashy	inflexible
speedy	bold	liberal
indecisive	harsh	sloppy
sweet	sour	bitter
intellectual	provincial	emotional
urbane	boring	sexy
momentary	puritanical	permanent
ordinary	savory	pungent

It is quite easy to come up with some of these. The authors have rendered possibilities for elegance, power, boredom, and speed (Fig. 154).

EXERCISE 5-4: Combining Marks

Tools and Materials: *Same as above*

Fig. 155 combines the matchstick marks and pen marks into a picture of a tree against the moon. Another student combined printed prefabricated marks into a rather effective evocation of light and sound (Fig. 156). Take some of the marks you devised or appropriated for Exercise 5–2 and combine them in a design.

FIGURE 155 A combination of pen lines and matchstick strokes

FIGURE 156 Combination of prefabricated patterns

EXERCISE 5-5: Paper Sculpture

Materials: *Construction paper*
Rubber cement
Transparent tape

Tools: *Ruler*
Single-edge razor blade
Metal straight edge
Scissors
Cutting surface

In Fig. 157 are paper sculptures done by students in a design class. They are nonobjective and functionless, but the principles involved in their creation can easily be used in making representational sculptures and devising packages. A number of the tricks entailed in building such constructions are revealed in Fig. 158. If you cut and fold a square of paper as indicated, you will be able to "crumple" it into a fat little ball made up of clustered points, as in Fig. 158B. If you make a second one, it is possible to glue the open points of it to the sealed form of the first (Fig. 158C); this will not only make it into a self-contained geometric solid, it will give the paper a remarkably strong structure. These "stars" make rather nice, and very inexpensive, ornaments. The main reason to do one, though, is to see the way paper can be folded into remarkably rigid structures. Notice that all of the vertical and horizontal folds go in a direction opposite to the diagonal folds. This is nearly always the case. The "waffled" and faceted shapes in Fig. 157 were all arrived at in this way.

Sheets of paper, metal, or anything that can be bent without breaking must change shape in two directions or be cut into if the surface is to be given a radically different configuration. The cylindrical

FIGURE 157 Paper sculptures by students

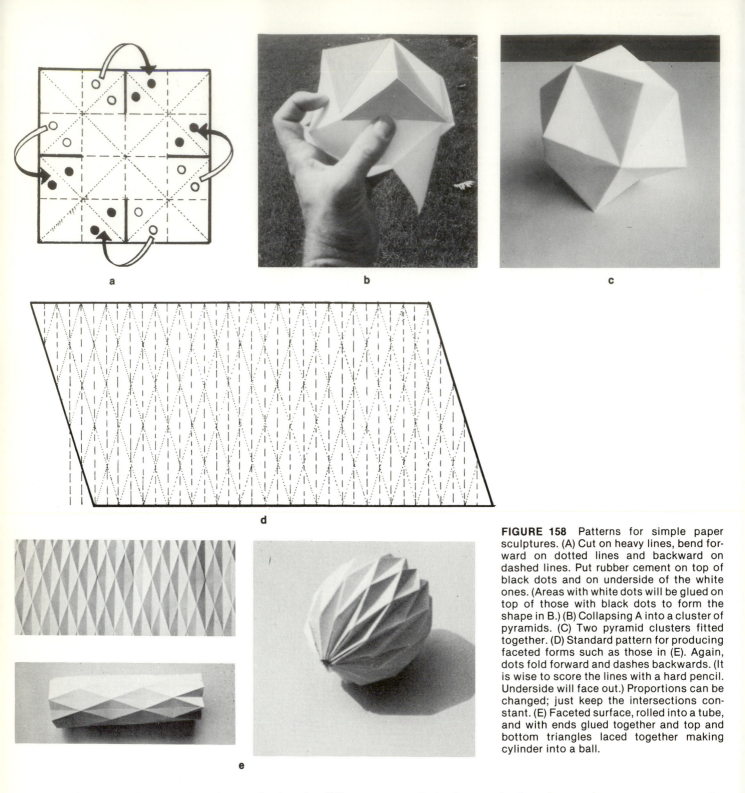

a

b

c

d

e

FIGURE 158 Patterns for simple paper sculptures. (A) Cut on heavy lines, bend forward on dotted lines and backward on dashed lines. Put rubber cement on top of black dots and on underside of the white ones. (Areas with white dots will be glued on top of those with black dots to form the shape in B.) (B) Collapsing A into a cluster of pyramids. (C) Two pyramid clusters fitted together. (D) Standard pattern for producing faceted forms such as those in (E). Again, dots fold forward and dashes backwards. (It is wise to score the lines with a hard pencil. Underside will face out.) Proportions can be changed; just keep the intersections constant. (E) Faceted surface, rolled into a tube, and with ends glued together and top and bottom triangles laced together making cylinder into a ball.

sculptures either consist of sets of tubes in different combinations or involve incisions and reverse curves. It is better to slice a tube and indent the strip than it is to remove a section because the former maintains some of the material that lends strength to the surface.

Also, when you multiply thicknesses, the sheet material becomes stronger. Thus, tubes of paper rolled from single sheets become enormously stronger. Accordion pleats close up a sheet into elongated ribs. A strip of construction paper that is 1 inch wide and 8 inches long is floppy and weak, but an $8'' \times 11''$ sheet pleated into eleven 1-inch accordion folds can support more than eleven $1'' \times 8''$ strips glued together would. The folds themselves add to the rigidity of the form.

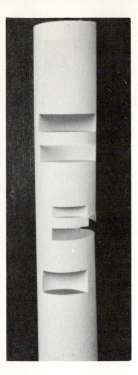

FIGURE 159 Smooth paper tube sculpture

FIGURE 160 A: "Blocking in" the outlines of the sculpture; B: Finished drawing

Create a paper sculpture using parallel horizontal and vertical folds in combination with parallel diagonals. Do another that uses only curves, cuts, and folds where one curve opposes another, as in Fig. 159.

EXERCISE 5–6: Three- and Two-Dimensional Forms

Materials: White drawing paper
Transparent tape

Tools: Single-edge razor blade
Drawing pencils (at least H, HB, B2, B4)

On a sheet of white drawing paper about 11″ × 18″ make three cuts toward, but not necessarily *to,* some corners of the rectangle. Then bend all four corners toward the center of the sheet, pinwheel fashion. Bend some over and some under if you wish; do whatever you think produces the most interesting shape. Attach the corners to the center with small pieces of tape.

Now use the resulting form as the subject of a pencil drawing. Sketch the outline very lightly and boldly (see Fig. 160A) and then lay in the lighter grays, progressing through darker pencils until you have reached an approximation of the light and shade patterns on the paper sculpture itself (Fig. 160B). There are two things to bear in mind when you un-

dertake to do this kind of shading: (1) After sharpening your pencils, stroke them over a coarse paper or *sanding block* so that they have a *chisel point* (Fig. 161), which will make it easier to achieve a broad, consistent tone. (2) When you use the pencils to apply areas of gray to the drawing, use a broad, polishing-like movement of the entire forearm and not a twitch-twitch-twitch of the fingers.

FIGURE 161 Pencil sharpened to a chisel point

FIGURE 162 A photograph "extended" with pencil drawing (A: the original photograph; B: the photograph plus the pencil extension)

EXERCISE 5-7: Matching the Machine

Materials: *White drawing paper*
A black and white photograph
Rubber cement
Tools: *Drawing pencils as above*

This is easy to explain but not so easy to do. Fig. 162 shows a photograph of some demolition rubble which has been extended with pencil drawing. The extension has nothing to do with what was actually beyond the perimeter of the photograph—as a matter of fact no horizon was visible because the photographer was shooting straight down—the artist has just made up things to prolong and stretch the original.

Take a photograph and trim the border from it. Mount it onto a sheet of white drawing paper. Then extend it, attempting to keep the grays as consistent with the photograph as possible.

EXERCISE 5-8: String Drawing

MATERIALS: *White drawing paper*
India ink
TOOLS: *Coarse string or twine*
Pens and brushes

FIGURE 163 A string-drawn "swan"

Dip the string into the ink. Then lay it down in a loop or two on a sheet of drawing paper you have folded in two and allow one end of the string to stick out so you can pull it without opening the paper. Press the top leaf down onto the bottom one and hold it down with your hand above the string. Using the free hand, pull the string out from between the sheets. Open the paper. The string leaves a wildly swirling set of forms. It is possible to turn these into other things. One student made a rather elegant swan (see Fig. 163).

notes for chapter 5

[1]Joy Kenseth, "Bernini's Borghese Sculptures: Another View," *Art Bulletin* (June 1981) Vol. 63, no. 2, p. 194.

[2]Ibid.

[3]Ibid.

chapter 6

color

There was a time when designers could ignore color theory and simply rely on printers, painters, photographers, weavers, and lighting experts to bring color relationships into some kind of correspondence with what they desired. To a large extent, that is the situation still. After all, the physics of color have little or nothing to do with aesthetic judgments about the use of color. Even where technical matters are concerned, instinct often overcomes some fundamental error.

Most artists do not use pure colors. As a matter of fact, they rarely speak of red or yellow paint but talk instead of cadmium red, alizarin crimson, and of cadmium, zinc, or naples yellow. Blues are ultramarine, phthalocyanine ("thalo"), cobalt, cerulean. There are also greens, violets, oranges, browns, and even varying kinds of whites, grays, and blacks. What is of major importance to a painter is the substance constituting the pigment in a given color and medium. About this we shall have a good deal to say later.

Designers need to know things about color theory that the old-fashioned painter does not. In the first place, designers are likely to be dealing with color photography and printer's inks. The first involves what used to be referred to in art classes as "light theory," only to be dismissed. The second is concerned with pigments and also with the effects of transparency and surface texture upon the appearance of those pigments. Let us deal with light effects first.

To most of us it seems odd, perhaps, that while the primary colors in pigment are (strictly speaking) yellow, magenta, and cyan blue, the primary colors in light are red, blue, and—green. Green? Yes, green. Speaking of light theory versus pigment theory is very misleading and awkward, however, for it divides a single phenomenon into what seem to be two dissimilar phenomena. Let's begin with the understanding that color does not exist in objects. Color exists in light. In terms of common-sense experience, color is almost a paradox; light rays are themselves invisible to the human eye and are therefore "colorless." However, a red ball (which is *in fact* colorless) appears to be red because it reflects red light rays. Furthermore, when you think you can see a "ray" of light, in the sense of a beam from a projector, for example, it is not the beam itself you see, but the dust particles or smoke or something else passing through the beam for it to illuminate.

When white light is broken down into its components by a glass prism (Colorplate 5), it produces a full range of spectrum colors: violet, blue, green, yellow, orange, red, and all of the intermediate gradations. White light striking white pigment reflects back almost entirely, but dull black paint absorbs nearly all of white light. Red paint absorbs all of the white light except red; the red is reflected back to our eyes and we therefore see the paint as a red area. Of course, this doesn't apply just to paint. It applies to everything. Lemons absorb everything but yellow, and the green felt on a billiard table reflects back nothing but green. (To ask what these substances "really" look like makes no sense. That is like wondering what a guitar sounds like apart from the sensations it produces in your ear, or asking what sugar tastes like apart from what it does to the taste buds in the mouth.)

subtractive color

The substances we have been talking about reflect a given color because they absorb the other colors, that is, they *subtract* from white light everything except red, or yellow, or green. This is easily understood if we think of the process in terms of white light passing through panes of glass that are, respectively, clear, red, yellow, and blue (Colorplates 6 and 7).

What is true of panes of glass is also true of red, blue, and yellow pigments. All pigments are substances which have been ground up into fine powder to be mixed with a vehicle such as linseed oil to make paint. Let us, then, turn our panes of glass into pigments by grinding them up. Clear window glass, finely ground, takes the form of a brilliant white powder. This is because the glass is colorless and therefore does not absorb any of the light but reflects all of it back to our eyes. Whiteness and clearness are merely different aspects of the same thing—the inability to absorb light. Why, then, isn't the powder transparent, like the sheet of glass from which it was ground? Because the tiny fragments are too irregular to allow light to pass through them. Even if these particles were perfect miniature panes of glass, they would pile up in opposition to one another, at random angles, so that more light would be reflected back than could ever pass through. Even large panes of glass lit by the sun behave like mirrors unless the viewer sees them at just the right angle.

The effect of white powder made from clear glass applies also to the ground glass pigments we have made from the other pieces of glass. The red pane becomes a red powder which absorbs blue and green light while reflecting red back to us. (See Colorplate 8.) Mixtures of the pigment powders will, of course, produce greens, oranges, and violets, among other colors.

As it happens, it would be impractical to make paint from ground glass. Colored glass is not nearly as intense as commercial pigments and would not be able to cover up what it was applied to. Ground glass in linseed oil would be similar to a very weak varnish.[1]

hue

Hue[2] is what most people think of when they hear the word "color." When you are arguing with someone about whether a garment you saw was green or blue-green, you are arguing about the hue. A simple way of describing hue would be to say that it represents a segment of the spectrum or colorwheel. In discussing color, artists often make reference to a *colorwheel* (Colorplate 9).[3]

You may have noticed that three of the hues are labeled with the numeral *1*, three with a *2*, and six with a *3*. The colors labeled *1* are called "primary hues." You cannot mix these from any other hues; you must begin with them. However, given these three primary hues, red, yellow, and blue, you can mix any of the others.[4]

If you mix two primaries together in equal amounts you will get one of the hues labeled *2*, hence their name, "secondary hues." Red + yellow = orange. Yellow + blue = green. Blue + red = violet. The hues labeled *3* are "tertiary hues," the mixtures standing between the primaries and the secondaries: red + orange = red-orange; orange + yellow = yellow-orange, and so forth. By adding a little more red to the red-orange one gets a redder red-orange, a scarlet. By adding quite a lot of red to a red-violet you can produce a red red-violet, a crimson. All such intervening colors are referred to as tertiary hues even though the number of variations is infinite. All of them are the result of combining any two of the three primaries.

hue combinations

Hues that are very similar, adjacent to each other on the colorwheel, are called *analogous hues*. Analogous relationships always involve a tertiary color since one of these is inevitably adjacent to any primary or secondary. When used together in a design, analogous hues tend to be harmonious and restful. Colorplate 10 has an analogous color scheme involving adjacent hues.

Colorplate 11 is very different from 10 in many ways, but one of the most striking differences is in the color organization. This is the opposite of analogous design; the hues are highly contrasting. The color scheme is *complementary,* involving hues that are directly opposite each other on the wheel.[5] The effect is one of excitement and high intensity. Colorplate 12 is a moderated version of that extreme, devised according to a harmonic principle known as "split complements." Such a combination of hues is usually arrived at in a very systematic way: for a given hue—say green—the hues adjacent to its complement (in this case, red-orange and red-violet) are appropriated for a design. A similarly prescriptive system commonly encountered in art and design is the so-called "triad" (Colorplate 13).

Rarely do artists work exclusively with such harmonic patterns as those defined by analogous, complementary, split-complementary, or triadic hues. More often they employ complex arrangements in which areas of one sort are played off against zones of another kind (see Colorplate 14).

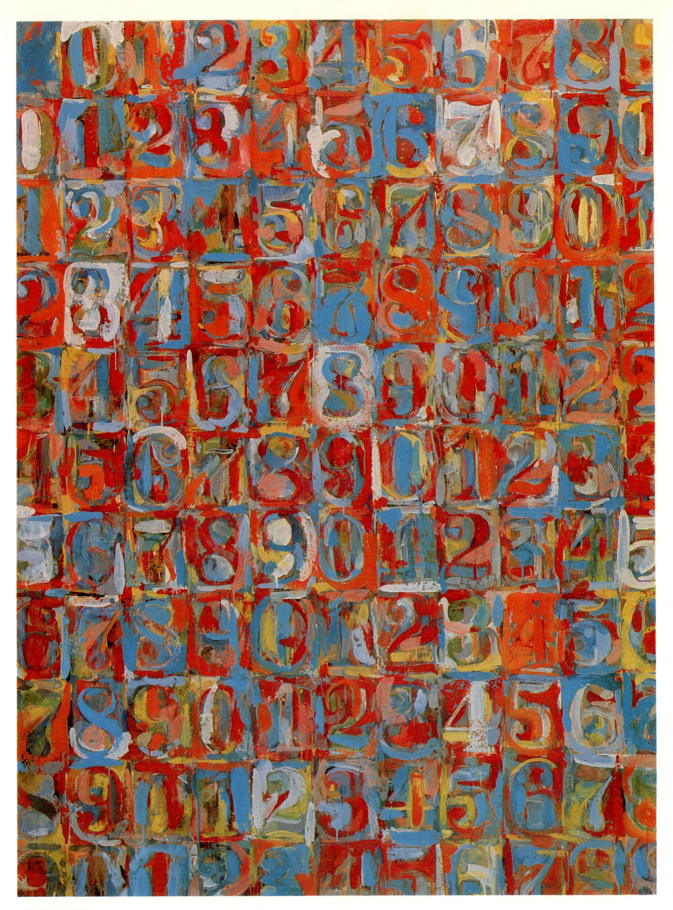

COLORPLATE 1
Jasper Johns. *Numbers in Color.* 1958–59.
Encaustic and collage on canvas, 5'7" × 4' ½".
Albright-Knox Art Gallery, Buffalo, N.Y.

COLORPLATE 2
Details from the "Super-graphs" (Figure 27)

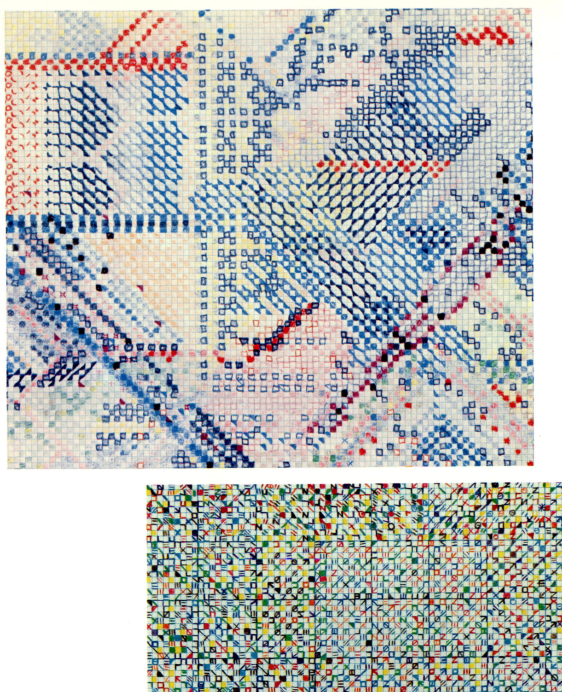

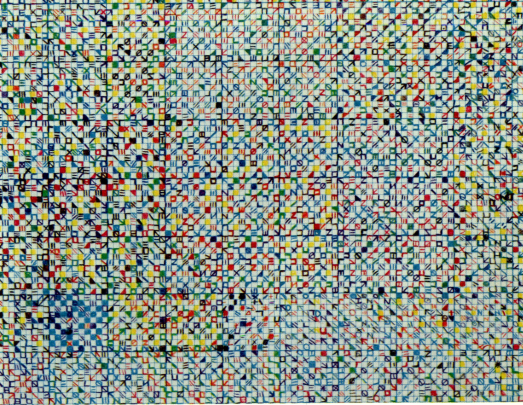

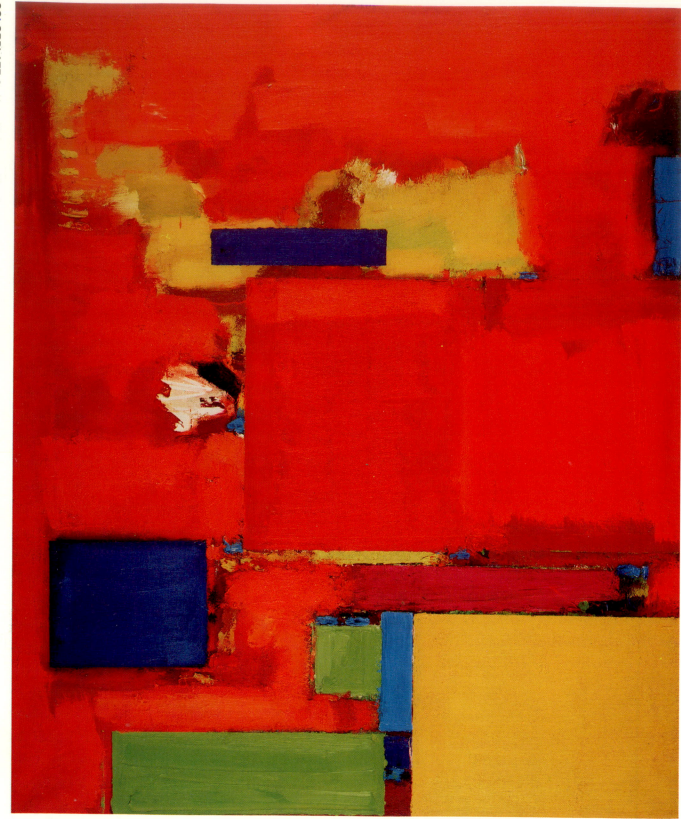

COLORPLATE 4
Stuart Davis. *Colonial Cubism*. 1954.
Oil on canvas, 45 × 60". Walker Art Center,
Minneapolis. Gift of the T. B. Walker Foundation.

COLORPLATE 5
Spectrum from white light broken down by a prism

COLORPLATE 6
Panes of glass casting colors onto a white surface

The clear glass permits the entire spectrum to pass through unimpeded and the white wall remains white. The red pane, on the other hand, permits only red light to pass through. It has not, as one tends to believe, "stained" the white light with redness; rather, it has subtracted everything except red. In the same way, yellow and blue panes absorb everything except yellow and blue, respectively.

COLORPLATE 7
Analytical diagram of Colorplate 6

In the examples from Colorplate 6, what colors were actually subtracted out? Since white light contains all colors of the spectrum in the form of red, blue, and green light intermixed, it should be easy to see that when red only is permitted to pass through a red pane of glass it means that green and blue have been absorbed. Blues passing through means the pane absorbs red and green. Yellow is perhaps a bit more difficult to deal with. Yellow glass subtracts out blue only, allowing red and green to pass through. Yellow light results from a mixture of red light rays with green ones, and a yellow pane of glass simply reverses the action of a blue pane.

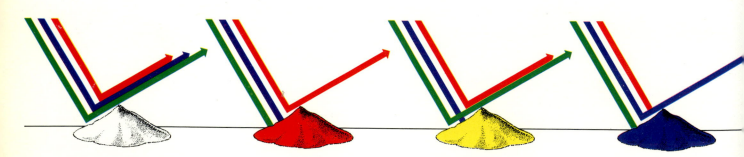

COLORPLATE 8
Light rays reflected from pigment powders

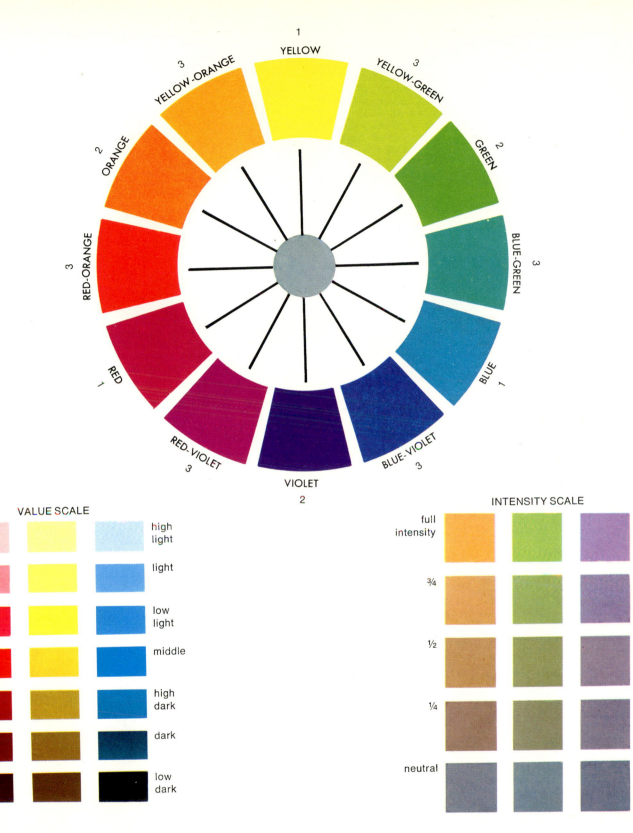

COLORPLATE 9
Colorwheel, value, intensity scales. (From John Adkins Richardson, *Art: The Way It Is* (Prentice-Hall and Harry N. Abrams, 1973).

First conceived of in 1660 by Sir Issac Newton, a colorwheel is nothing more than the hues of the spectrum bent around into a circle. (There is, however, one color in the wheel that does not appear in the rainbow a prism gives us. Red violet is used to tie the ends of the spectrum together in a circle.) It would be possible to make the circle of hues continuously graduated into one another, but it is customary to divide colorwheels into twelve units. See footnote 3 of this chapter.

COLORPLATE 10
Mark Rothko. *Orange and Yellow*. 1956.
Oil on canvas, 91 × 71″.
Albright-Knox Art Gallery, Buffalo, New York.
Gift of Seymour H. Knox

COLORPLATE 11
Michael Beuse. *Counterchange #2*. 1969.
Serigraph, 17 × 17″. Private Collection

COLORPLATE 12
Split complements.
Design after Michael Beuse, *Counterchange #2*

COLORPLATE 13
Three different triads.
Designs after Michael Beuse, *Counterchange #2*

These show us the same forms colored according to a triadic harmony in three different ways. *A* is the obvious one; it is red, yellow, and blue, the primary triad. *B* is also based on an equilateral triangle but shifted over one hue to yellow-green, red-orange, and blue-violet. *C* takes three other hues equidistant from one another on the wheel: orange, green, and violet.

A

B

C

COLORPLATE 14
Pablo Picasso. *Girl Before a Mirror*. 1932.
Oil on canvas, 63 ¾ × 51 ¼". The Museum of Modern Art,
New York, Gift of Mrs. Simon R. Guggenheim.

In this Picasso the mirror image is made up mostly of analogous hues while the background and the girl herself are constituted of complementary and triadic contrasts. *Girl Before a Mirror* is a particularly bold and well-orchestrated example of such color balance but you will find some hint of this sort of equilibrium in many fine paintings.

A

COLORPLATE 15
A: Additive mixing of red, yellow, blue
B: Additive mixing of red, green, blue

Whereas the mixing of the pigment primaries magenta, cyan, and yellow will produce an indiscriminate "gray"—even if light beams are used instead of paint—light primaries intermixed produce pure white light. In the diagram a red (magenta) beam, a blue (not cyan), and a yellow overlap and make a weak brown. However, the overlap of light beams of the primary colors of light—red, blue, and green—makes white. The secondary mixtures of the lights results in hues that are the same as the primaries for pigment.

B

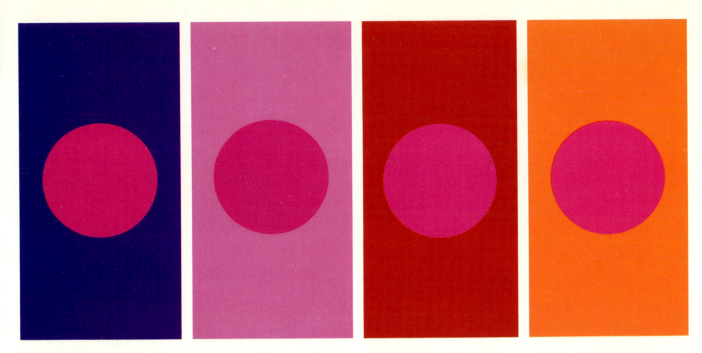

COLORPLATE 16
Simultaneous contrast

The circles are all identical, but they appear to be different because of the operation of simultaneous contrast. The result is subtle but the principle involved is quite simple. A black put against a white looks darker than when it rests against gray. Moreover, the white looks whiter with the black against it. Each gives its opposite, black or white, to the other. In a similar way the discs communicate their opposite attributes to whatever background they are against and at the same time the backgrounds impose their opposites upon the discs.

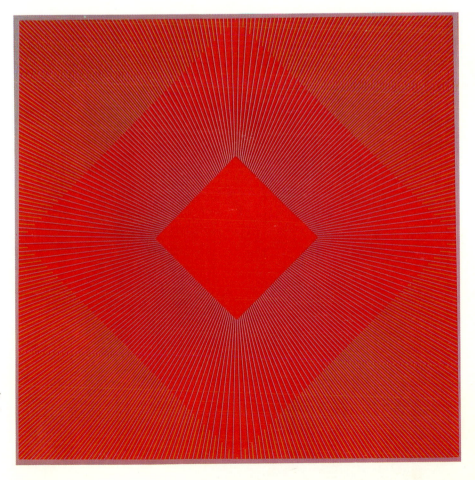

COLORPLATE 17
Richard Anuszkeiwicz. *Splendor of Red.*
1965.
Acrylic on canvas, 72 × 72″. Yale University
Art Gallery, New Haven. Gift of Seymour H.
Knox.

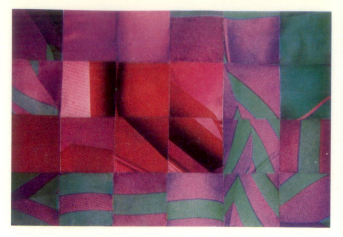

COLORPLATE 18
Color montages by students

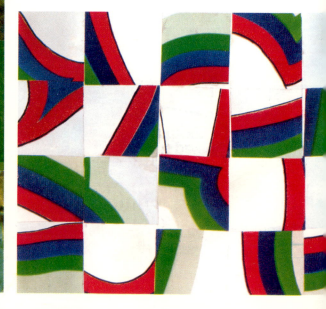

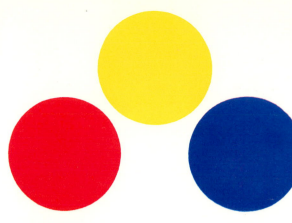

COLORPLATE 19
The primary hues as they were once conceived.

COLORPLATE 20
Red paint samples

COLORPLATE 21
Stephen Posen. *Variation on a Millstone.* 1976.
Oil on canvas, 86 ½ × 68 ½".
Pennsylvania Academy of Fine Arts, Philadelphia

Variations on a Millstone is even more puzzling in its appearance than in its title because it combines an illusionistic treatment of texture with what at first seems to be a contradictory treatment of the subject so far as color and space are concerned. It is like some sort of puzzle. Even when one knows the "solution," it is impossible to dismiss completely the disconcerting aspect of Posen's fabrication.

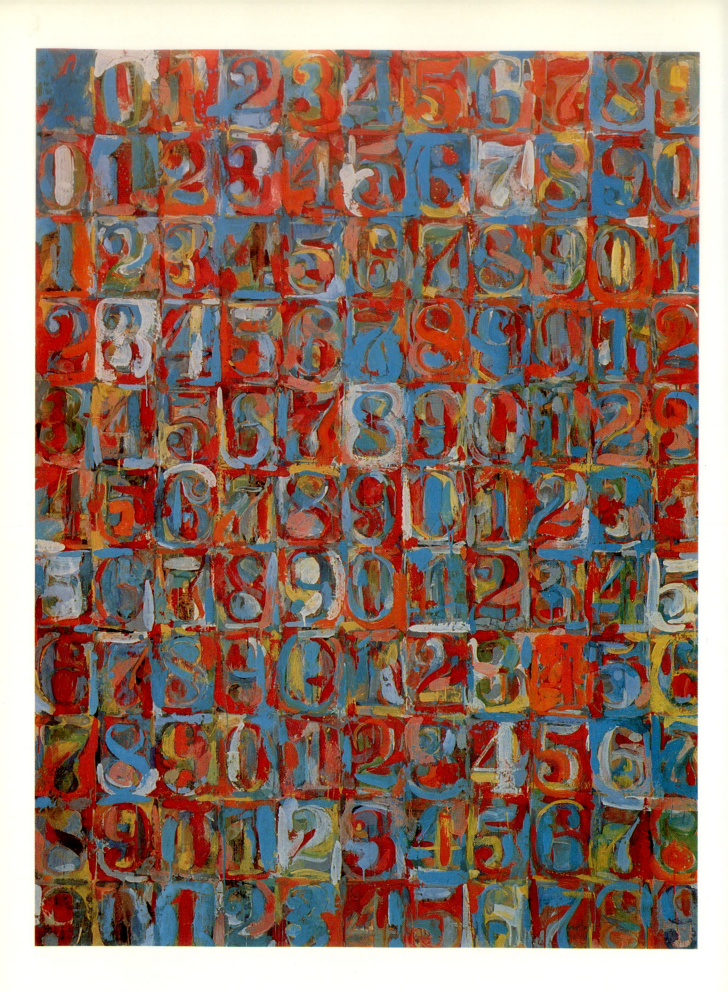

| yellow | magenta | cyan | black |

| yellow | yellow and magenta | yellow, magenta, and cyan | yellow, magenta, cyan, and black |

COLORPLATE 22
Process color separation

The process camera is used to take four photographs of a piece of color art through four differently colored filters. The filters are the exact complements of the process primaries—yellow, cyan, and magenta—plus one yellow filter. When the full color picture is photographed onto black and white film through a violet filter, the yellow components of the full color image are translated into dense blacks and dark grays while the other colors are bleached out so that the film doesn't see them at all. A green filter turns reds to blacks but drops out yellows and blues. An orange filter brings out the cyan blue but suppresses reds and yellows. Finally, the yellow filter is used to balance the black and white values throughout the image. Each of the four film images produced in this way is black and white (that is, black and transparent), just like any film negative. From the negatives four separate printing plates are prepared. When these are run inked with the appropriate hues (magenta, cyan, and yellow) and black, the superimposition of all four will give you close approximation of the color range in the original art. It is typical to use a different screen (such as 120 line) for the yellow printer than for the other three colors (100 line) to avoid moire patterns of the kind shown in Figure 318.

DENSITY OF SCREEN
0% 10% 40% 100%

ANGLE OF DOTS
60° 45° 105°

Y = YELLOW   65 LINE
B = BLUE   SCREEN
R = RED

COLORPLATE 23A
Mechanical color separation

This display of colors was entirely rendered in black and white with india ink and shading sheets on transparent overlay films. (The panel giving the screen densities shows what the benday dot patterns look like to the person using them.) Printing plates made from the overlays and inked with transparent process inks (yellow for one plate, red for another, blue for a third, and black for the fourth) are run through a press so that their impressions are superimposed on one another, producing the combinations here.

COLORPLATE 23B

Special overlay films and related media that make color separations of this kind far simpler to do are available from Ohio Graphic Arts Systems, Inc., 26055 Emery Road, Warrensville Heights, Ohio 44128 under the trade name Grafix. The sample shown here combines 65 line screen densities of 0, 25, 50, and 100 percent.

COLORPLATE 24
Two-color designs employing non-process colors for decorative effect

temperature of color

Another feature of color is the relative warmth or coolness of hues. The identification of reds, yellows, and oranges as warm colors, and blues, blue-greens, and violets as cool colors, no doubt derives from our experience of nature. Extremely hot things are usually red or orange and the ocean, sky, distant mountains, and ice cubes are bluish. Artists have come to refer to hues lying toward the red-orange side of the colorwheel as "warm" and those lying toward the blue-violet end as "cool." Within the various hues, too, there are different temperatures. Thus, a red-violet is a cooler red than a red-orange, and a yellow-green is warmer than a green. The connection between hues and actual temperature is rather loose—as observe "white hot" and blue flame—and yet most people will perceive the temperature of a room painted blue as lower than it actually is and will think a room painted warm brown is warmer than it is.

value

Value refers to the lightness or darkness of a color. The hue of a color depends upon the *kinds* of light rays the color absorbs from white light and upon which ones it reflects back. The value of a color depends on how *much* of the light is absorbed. White reflects most of it back, black absorbs most. Pure violet absorbs far more light than pure yellow. If we wish, however, we can lighten the violet to the point where it is as reflective as the yellow by introducing white into it. When this is done, the resulting color, orchid, is a *tint* of violet. A tint of a hue is a lightened version of the hue.

There are various ways of darkening the yellow to make it more light-absorbent. We could add black to it, for instance. A yellow so darkened would be called, technically, a *shade* of yellow. (Many decorators use the terms *tint* and *shade* interchangeably. Color theorists don't because it is obviously useful to have a simple linguistic distinction of this kind.) In Colorplate 9 a value scale gives some indication of possible comparisons and contrasts among hues. Needless to say, the lightest of all colors is white and the darkest is black. (For our purposes, it is convenient to call these colors even though they are hueless. Other writers would think of them as colorless values.)

Darkness or lightness of value has an important bearing on the character of any design. Works of art that are highly dramatic, such as Gérôme's *Police Verso* (Fig. 101A) or *The Third of May* (Fig. 164) by Francisco Goya (1746–1828), usually contain areas of high contrast. In the latter work, which represents French troops crushing a rebellion by the people of Madrid, the principal victim is in the center of a veritable explosion of light. The effect is all the more striking because his flesh is dark. He throws out his arms in a gesture of defiance and signifies that he, like Christ, is being "crucified." The firing squad is shadowy and featureless; the soldiers have no personalities; they are merely instrumentalities of French imperialism, the tools of the oppressor, Napoleon.

FIGURE 164 FRANCISCO GOYA. *The Third of May, 1808.* 1814–15. Oil on canvas, 8′8¾″ × 11′3⅞″. The Prado, Madrid

Such extreme contrasts of dark and light make for a highly charged impact. That is true also when they occur in graphic designs. The travel brochure cover in Fig. 165A is emphatic; the lights stand out against the blackness like shouts in the night. Figs. 165B and 165C, by contrast, are arresting because of their subtlety. The first hints at soft nights on the canals and the second makes one think of luminous, sunny afternoons in the city of the Doges. J. W. M. Turner (1775–1851) did paintings of Venice that are so light as to approximate luminous phantoms on canvas (see Fig. 166). Henry Tanner (1859–1937), the first internationally successful black artist, painted *Abraham's Oak* (Fig. 167) in middle-range values and muted hues, adding a note of mysterious religiosity to what would otherwise be a rather ordinary landscape. Ad Reinhardt (1913–1967) did abstractions so dark that color changes within them are virtually imperceptible until after several minutes of intense concentration. Reproductions (see Fig. 168) usually do not convey the magic of the emergence of somber rectangles from sunless depths because the photographs always either exaggerate the contrasts or make them invisible.

Works in which the majority of values are, like those in Figs. 165C and 166, lighter than middle gray are called *high-key.* Those in which the tones fall below a middle gray, as in Figs. 165B and 168, are referred to as *low-key.*

One must understand that high-key paintings or other designs may contain low-key areas, and some largely dark works will have bright spots. The Turner does not seem pale, exactly; it looks brightly lighted, as if filled with some glowing mist. This painter was a master of *luminosity,* but the principle he uses to effect an impression of luminous brilliance is

FIGURE 165 A design for a travel brochure cover. A: high contrast; B: Low key; C: High key

a

b

c

FIGURE 166 JOSEPH MALLORD WILLIAM TURNER. *The Dogano, San Giorgio Maggiore, le Zitelle, from the Steps of the Europa.* 1842. Oil on canvas, 24 1/2 × 36 1/2″. The Tate Gallery, London

FIGURE 167 HENRY TANNER. *Abraham's Oak.* c. 1897. Oil on canvas, 21 1/2 × 28 3/8″. National Museum of African Art, Eliot Elisofon Archives, Smithsonian Institution, Washington, D.C.

FIGURE 168
AD REINHARDT. *Abstract Painting, Blue.* 1952. Oil on canvas, 75 × 28″. Museum of Art, Carnegie Institute, Pittsburgh. Gift of the Women's Committee, Museum of Art, 1965

FIGURE 169 JAN VAN EYCK. *Madonna of the Chancellor Rolin.* c. 1433–34. Oil on panel, 26 × 24⅜″. The Louvre, Paris

easy to explain. It is achieved by contrasting intense darks with a field made up of highly contrasting hues (such as blue and yellow) which have been tinted to give them the same value at a very high key.

If you look at a lot of paintings, you will eventually realize that works that excel in the representation of minuscule detail are nearly always in the lower value register. The van Eyck (Fig. 169) and the Rembrandt (Fig. 170) are excellent examples and both are predominately low-key pictures. The reason for the association of darkness with detail is that it is possible to achieve more variation within and among highlights in a dark painting than in one that is high-key.

FIGURE 170 REMBRANDT VAN RIJN. *The Golden Helmet.* Gemaldegalerie, Staatliche Museen Preubischer Kulturbesitz, West Berlin. Photo: Jörg P. Anders.

Rembrandt van Rijn (1606–1669), perhaps the greatest name in Dutch art, cast his brother, Adrian, in gloomy shade when he did *The Golden Helmet* but marvelous highlights describe the intricate jeweler's chasing of the helmet and the luster of the cuirass the tired old warrior wears. In *The Golden Helmet* a middle value appears to be quite light (far lighter than it would seem against white) because it is seen against a very dark background. Therefore, Rembrandt was able to enhance the impression of complete fidelity to nature with *additional* highlights of pale yellow, cream, and ivory.

intensity

Of all the attributes of color the most difficult to explain is intensity. *Intensity* is the term used to describe the brightness or dullness of a color and there is a tendency for people to confuse intensity with value. If you look at the intensity scale in Colorplate 9, you will see that it has to do with how much of the hue is present in a given color sample. *How* red is this red? How green is a given green?

Anything you do to a color that does not change its hue will vary its intensity. If you mix yellow into a pure red, you modify the hue; the red turns into red-orange or orange. If, however, you add black to the red, you change the shade of the red and also its intensity. It becomes less bright. Tinting the red with white makes it a pink but also weakens it as a red, lowering its intensity or purity or brightness, even while making it lighter. However, neither of the last acts—tinting or shading—changed the hue from red. Odd though it may seem, the introduction of green into the red will also make the red duller without changing its hue.

Look at Colorplate 9 again. Notice that the colors at the top of the intensity scale are quite bright and that they decrease in brightness as they go down. The *value* of all those squares is the same; only the brightness has changed. The bottom square in all three rows is a "neutral," a gray. The neutral in the left column is the dullest orange it is possible to have. Similarly, the last square in the second row is as dull a green as you can get. In the same way, the lower-right-hand square is the dullest of all violets. Also the gray squares are, respectively, the dullest blue, red, and yellow possible. In fact, the neutral in the left-hand column is the product of an equal mixture of orange and blue (tinted to maintain the value of the original orange). Likewise, the neutral in the next row is a mixture of green and red, and in the right row it is a mixture of yellow and violet. Those particular hues neutralize each other when they are mixed. A little bit of a complement dulls slightly, a large amount cancels out completely the opposite hue with which it has been mixed.

Obviously, a neutral can also be produced in other ways than by mixing two complements. The three primaries can be combined or black and white can be mixed. However, one can obtain far richer grays—warm, cool, soft, or clear—by mixing complements and adding white. Black tends to deaden colors as well as darken them and it is wise to use it very sparingly indeed.

The usual way of dulling a color is to mix into that hue its complement. Since shadows on objects are not only darker than lighted areas but also duller, it is typical for realistic painters to paint them by dulling the color of the object with a bit of the complement. In other words, the shadow on a green vase will be green with a good deal of red mixed into it. What this produces in lay terms is frequently a "brown." If you study the intensity scale in Colorplate 9, you can observe that brown is nothing but a step on the way to the neutral.

additive color

Let's consider again our panes of colored glass. Pretend that these are projector lenses of yellow, red, and blue in powerful lamps and that we have extra lenses (or filters) which can be inserted at will. Of course, each lamp can project a spot of colored light onto the white wall (Colorplate 15A). What if we want an area of white light on the wall in the same spot? If we overlap all three of the light beams, the result will be a rather dim, brownish hue. If, however, we slip a blue lens in with the yellow one in the third projector, it will throw a spot of green light onto the panel. Then, the overlap of the red, blue, and green beams will produce pure white light (Colorplate 15B). This is the *additive* mixing of color.

> Blue and red together = magenta
> Blue and green together = cyan
> Green and red together = yellow

In other words, the secondary mixtures of light are the same as the primaries for pigment. This may be surprising to someone who first learns of it. The reaction indicates only that he or she is not analytical in what might be called a "logical" fashion. After all, we have been using the three projectors in Colorplate 15 to *add* together what we *subtracted* in Colorplates 6 and 7. It's embarrassingly obvious. Yet, most of us do not think about sensation in this way.

The most familiar application of additive color in a direct way is the color television screen. It is made up of tiny phosphor dots. A magnifying glass held up to the screen will reveal these dots grouped in sets of one red, one blue, and one green. When all three dots are illuminated, the result is white. When the dots are not stimulated at all, black results. Other colors result from stimulation of individual primaries alone or in combination. Additional colors can be evoked by varying the intensity of the light beams. Thus, red coupled with a very weak green produces a low-intensity yellow—a brown. Pink is white light in which the red is a trifle stronger than the blue and green. Lavender is similar, but the green is weak. And so on.

Actually, additive color mixing can be done with *any* three hues, so long as each falls in a separate third of the spectrum, but green, red, and blue will produce the widest range of color.

the color of adjacency

The setting in which a given hue resides has a very important bearing upon its appearance and effect. In fact, there is no circumstance in which the colors next to one another are without an effect upon one another; that is because of the phenomenon known as *simultaneous contrast,* the tendency of a color to induce its opposite in hue, value, and intensity in an adjacent color (Colorplate 16).

In the same way, hues appear at their most brilliant when complements are in conjunction; red makes green more green and violet makes yellow yellower. Orange is at its most orange when seen against blue. That, of course, is because the hues lend their complements to the color next to them, intensifying their complements still more. However, a yellow will also make a "true" blue seem slightly violet. A red will make it look more like cyan by contributing a tinge of green to it. A pure, bright green will make the blue seem less pure, because the complement of green is red.

So called Op Artists have used this principle more than any other, perhaps. Richard Anuszkiewicz (b. 1930) depended upon simultaneous contrast for the effect in the work reproduced as Colorplate 17. The background color is the same throughout; the lines of white and yellow make it appear different.

exercises for chapter 6

Before diving headlong into studio projects, it may be wise to deal with some practical problems. For instance, in theory, any mixture of equal amounts of two primary colors results in a secondary. Thus, blue + yellow = green, and yellow + red = orange. Quite true, but that green or orange will not be the brightest or most intense version of the hue. The theory is not wrong, but the pigment facts do not correspond to theoretical truths. A pigment that reflects blue can be mixed with one that reflects yellow to yield a green, but that green will only approximate the green labeled *2* on our colorwheel. That is because a pigment which, by its nature, absorbs everything except green will reflect a brighter, purer green than can be mixed from pigments which, by their nature, reflect blue and yellow. In fact, if you wish to make the pure green resemble a mixture of blue and yellow, you can put a little red in it. Pigment mixtures in fact do not accord with theory nearly so well as mixtures of light will. Therefore, a painter, illustrator, or designer should have on hand many cans or tubes of different hues and pigments in order to have available secondaries and tertiaries of optimum intensity.

Also, there is the matter of the *staining power* of each hue. It doesn't take the addition of much primary blue to change greatly the hue of most other colors because blue is not only the coolest of the three primaries, to most people it also appears psychologically to be the darkest, although magenta in a pure state is at least as light-absorbent and, in fact, looks blacker than black to a process camera. Yellow, on the other hand, is clearly the lightest of the primaries and is therefore the weakest in staining power. Yellow and blue behave much like black and white. It takes an enormous amount of white to change a black perceptibly, but a few drops of black will alter white quite noticeably. Similarly, a pinch of blue turns a large mass of yellow green; to turn blue into something that is even slightly green requires a lot of yellow.

Look at Colorplate 20 and identify the hues as best you can. They appear to have little in common apart from a general redness. The fact is that they are all the same color, cadmium red medium, according to the labels on the tubes from which they were squeezed. Colors vary quite dramatically from one manufacturer to another. Therefore, most of us divide our loyalties among different companies, depending upon what uses we make of the available pigments. A given artist might buy raw umber from Grumbacher, another brown from Shiva or Permanent Pigments, and reds from Winsor-Newton or Boccour in order to secure the palette of colors desired.

At this point in your development as an artist or designer the easiest colors to mix are *designer colors,* which are prepared for the specific purpose of achieving the brightest, flattest, most easily and predictably mixable hues. Among the problems these paints present is that transparencies are not easy to attain; they appear flat and dull compared with those of watercolors. Too, the pigments incorporated together to produce the magentas, cyans, and their perfect secondaries and tertiaries are not sufficiently permanent for work intended to last indefinitely. That eliminates them from the fine arts. However, they are useful for designs that are intended to be photographed and reproduced and for which the original artwork is only a starting point.

EXERCISE 6-1: Color Montage

Materials: *White drawing paper*
*Picture magazines containing plenty of
color reproductions*
Rubber cement

Tools: *Single-edge razor blade or frisket knife*
Ruler with a metal edge
Cutting surface

Picture magazines such as *Life, The Smithsonian,* sports, fashion, and homemaking magazines, printed calendars, and mail-order catalogues, with their richly hued photos and designs and their varied and enticing textures and patterns, are a practical, ready-made source to accelerate the practice of combining colors.

Do four separate designs using square or rectangular scraps of the same size and shape throughout, working within a grid constituted of 1-inch-square modules. Each of the four designs should employ at least one set of characteristics from each of the three groups listed below and no design should use any characteristic that has been used in another one. (See Colorplate 18 for some examples.)

A. *Color Harmonies*
   1. Triadic hues
   2. Analogous hues
   3. Complementary hues
   4. Split complementary hues
   5. Complex hue pattern
B. *Scale (exterior dimensions)*
   1. 1″ × 2″
   2. 3″ × 5″
   3. 6″ × 6″
   4. 8″ × 10″
   5. 2″ × 15″
C. *Surface effects*
   1. Flat, solid colors
   2. Prominently textured areas
   3. Patterns
   4. Fragments of reality
   5. Fragments of art

EXERCISE 6-2: Collage of Painted "Chips"

Materials: *White drawing paper*
Waterbase paint (designer colors, acrylics, etc.) having a full range of hues
Rubber cement

Tools: *Red sable watercolor or lettering brushes*
T-square
Triangles

Ruler with a metal edge
Single-edge razor blade or frisket knife

Using flat, solid zones of color, paint sixteen separate areas slightly over one inch square. Four of these should be in a primary or secondary hue (such as yellow) and four in some contrasting hue (such as blue, blue-violet, green, or blue-green). For purposes of discussion, let us say four areas are yellow and four are blue-green. Each set of four should represent its hue as high-light, low-light, high-dark, and low-dark values.

Next do two more sets of four areas. Each of these four areas should preserve the same value relationships as the first two sets but should mix the two contrasting hues. In one set the first hue (in our example, yellow) will predominate, in the other set the contrasting hue (blue-green) will predominate. This should give you something of the following kind: one set of four yellow areas, going from high-light through progressive steps to low-dark, a second set of yellow-green areas doing the same, a third set of green areas matched to the values of the other two, and a fourth set of blue-green areas keyed to all of the others so far as value is concerned.

In painting the areas, use the following technique: Paint a spot slightly larger than the one inch square and do not concern yourself with a neat edge. Apply the paint in at least two rather thin coats rather than a single heavy one. Mix the color in a pan or dish in small amounts. (Use a red sable brush; nothing else will do a proper job.) Rinse out the brush frequently.

When the spots of color have dried, measure off 1-inch squares and cut them out neatly, using the razor blade or knife along the ruler's edge. Mount these *chips,* as color samples are called, in a rectangle 4 by 4 inches square so that the high-lights are on the left, the low-darks on the right, and the intervening values between. In the same way, place the contrasting hues at the top and bottom registers of the rectangle with their mixtures in between. See Fig. 171 for a diagram of our yellow/blue-green example.

FIGURE 171 A diagram of Exercise 6-2, using yellow and blue-green as extremes.

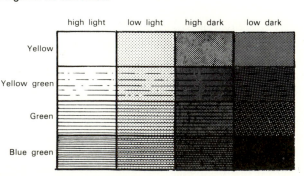

EXERCISE 6–3: A Study in Close Values

Materials: White drawing paper
Picture magazines
Rubber cement

Tools: Single-edge razor blade or frisket knife
HB, B, or 2B pencil

It should seem obvious that the standard hues of the colorwheel, seen at pure intensity, have values different from one another. In practice, however, taking for granted that blues are dark and yellows light has been the downfall of many aspiring artists and designers. To work in a very confined range, as Turner, Tanner, and Reinhardt did (Figs. 165, 167, 168), is unusual, but doing so can prove instructive.

Select a good color reproduction of a painting by a renowned artist. It should be in a normal key range, having darks, lights, and middle values as do, for instance, the Michelangelo (Fig. 141B), Gérômes (Figs. 100A, 101A), or Holbein (Fig. 32). Make a relatively simple contour (outline) drawing of the elements of the painting, concentrating on simplifying them into uncomplicated shapes (see Fig. 172). You can be more intricate in your rendering or less. The idea is to simplify things to make for easy cutting while at the same time arriving at pleasing shapes. This rendering should not be conceived of as copying; think of it as a reinterpretation of the original, inevitably affected by the nature and the limitations of the materials you are using.

FIGURE 172 Reduction of Michelangelo's *The Holy Family* (Fig. 141B) to outline pattern.

Once the outlines have been established, cut out pieces of paper from the magazines and fit them to the outlines you have drawn. You can use your own discretion about what hues to use; you needn't follow the scheme of the original. There is but one rule:

***all of the colors used should be as
close in value as possible.***

All of them can be extremely high-key, all within middle value range, or all low-key.

Most of us find this a very challenging and valuable project. Among other things, it helps teach one to *squint*. Squint? That's right. Painters, print-makers, and designers are always squinting because a narrow squint blurs details and reduces things to their primary value relationships. Landscape painters squint at the world in order to simplify nature and reduce it to masses of color; designers squint at blocks of type to see the overall effect; critics squint at art works to resolve complexities into patterns. Squinting at your close-value study will void some of the hue and intensity of the color patches and make value similarities and differences more readily evident.

EXERCISE 6–4: Lantern Slides and "Found Color"

Materials: Ten pieces of 3 1/4″ × 4″ projector slide cover glass (available from Kodak)
Masking tape (1/4″ or 1/2″ width)

Tools: Principally a Slide King projector
Various transparent and translucent materials, such as candy wrappers, inks, stains, dyes, cellophane, wax, netting, transfer type, and gelatines

This exercise deals with organic and geometric shapes, with visual organization, and most important, with color luminescence.

Luminosity, as a first-hand visual effect rather than a pictorial illusion, has to do with relative transparency under illumination. When a color is "opaque," it cannot be seen through. Titanium white oil paint, applied evenly, as it comes from a tube, obliterates whatever is beneath it; it's opaque. That same paint applied to a pane of glass will allow no light to pass through it. Something is "translucent" when it is not completely opaque yet does not permit light to pass through unimpeded. Frosted glass, milk glass, was paper, and opalescent stained glass are examples of translucent substances. When light is allowed to pass through a substance more or less completely, even though some colors may be subtracted out, that substance is called "transparent." Spotlight gelatines, clear stained glass, and watercolor washes are transparent.

For this exercise make five slides, using ten pieces of 3 1/4″ × 4″ projector slide cover glass. Arrange various transparent and translucent materials into designs which will be "sandwiched" between two plates of the glass. The plates will be joined together by a "frame" of masking tape.

Do not try to put too much material between the plates. These slides must be kept flat enough to fit into a projector. Any material is acceptable so long as it is reasonably thin, will not damage the projector, and is not extremely flammable. Do not use jam (which molds and runs when the glass gets hot), petroleum jelly (which becomes soggy), or cooking oil (which runs down and out).

All of the designs will, of course, be dependent on subtractive coloration. It is interesting to use two or more projectors and overlap the images of two or more slide experiments because this will frequently produce additive effects.

EXERCISE 6-5: Simultaneous Contrast

Materials and Tools: Same as for Exercise 6-2

Create some effects of the kind shown in Colorplate 18. It might, in this connection, be interesting to do in color something like our study in values in Fig. 173 in which the central strip remains a constant gray. It *seems* to change only because its relationship to the background is inconstant.

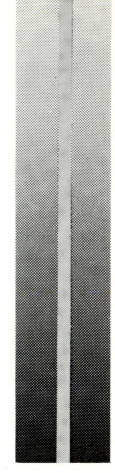

FIGURE 173 Gray strip against a graduated background

notes for chapter 6

[1] To have much covering power a pigment must have a higher refraction index than the vehicle in which it is suspended. *Refraction index* is the technical term for the relative change in velocity of light caused by its deflection when passing through a substance. (In other words, how much does it slow down?) The refraction index of boiled linseed oil is in the neighborhood of 1.5. If the particles mixed into it have a similar or lower refraction index, it will be as if the particles were absent except for an alteration of the texture and viscosity of the oil. Plain window glass, finely ground, has a lower index than linseed oil. When mixed with the oil the white powder produces a paint with no more opacity than clear varnish. Titanium dioxide, however, has a refraction index of 2.76; titanium white oil pigment is one of the least transparent of all paints and has extremely high covering power. Lead white (sometimes called "flake white") is as opaque as titanium white but has the disadvantage of less permanence; over the years it yellows more than titanium. Zinc white is a clearer white than either but it has less covering power.

What is true of the white pigments also holds for pigments of other hues. Each has its own peculiar attributes.

Paints that are *impermanent* are subject to change in appearance by yellowing, darkening, or becoming dull over a period of time. This is an important consideration in the creation of a painting intended to be seen by future generations and purchased as a long-term investment, but permanence is given little consideration by graphic designers doing posters that will soon be torn down and thrown away. Contrariwise, the capacity of a given pigment to reproduce a desired color accurately is a major consideration for the designer and is of no concern to the painter of portraits.

[2] While the terminology employed here is generally understood by artists and designers, printers use slightly different terms for color characteristics. Thus, what most laypeople call "the color" and artists called "hue" (that is, red, green, red-violet, yellow-green, orange, etc.) the printer refers to as *chroma*. Darkness or lightness, which most of us refer to as "value" or "tone," is called *gray* in a printing plant. And intensity of hue—the brightness or dullness of a color—is named *strength* by commercial printers.

[3] We have chosen the most commonplace kind of colorwheel based on three primaries (process red, blue, and yellow) placed equidistant from one another and have appropriated the three-dimensional terminology (hue, value, intensity) used by theorist Arthur Pope in *The Language of Drawing and Painting* (Cambridge: Harvard University Press, 1949). There are other widely used theories, which vary significantly. For example, the Ostwald colorwheel (c. 1916) accepts four primary colors (red, blue, green, yellow) equidistant from one another on the circle. Albert H. Munsell's 1898 colorwheel takes into account what appear to be five primaries: red, violet, blue, green, and yellow. Munsell began with the physicists' light primaries (green, red, blue) and then planned the wheel so that any three equidistant points on the wheel work out. In both Ostwald's system and Munsell's the configuration of complements changes radically. For Ostwald the complement of yellow is ultramarine blue, for Munsell it is blue-violet. A case can be made for tremendous range in what might be called a complement. In point of fact, once we have accepted the process colors for our primaries, the complements we designate turn out to be virtually identical to those of the other systems; our violet is close to what Ostwald calls "ultramarine" and Munsell "blue-violet."

That is not to say that only the words are different; there are other differences. However, none of the theories is "right" nor "wrong"; they can all work.

[4] For years painters considered the hues in Colorplate 19 to be fair approximations of the ideal primary colors from which all others (except black or white) could be mixed. Later investigators know that the real primaries for pigment are those labeled with a *1* in Colorplate 9: magenta, cyan (turquoise), and yellow. What most of us think of as a "true red" can be evoked by combining a magenta with a little yellow. Putting a touch of magenta in cyan creates the kind of blue that isn't blue-greenish. In the world of graphic design, however, the pigment primaries are known as "process red," "process blue," and "process yellow."

[5] Complements are not merely what are across from one another. The wheel is merely designed to achieve that disposition. If you are in doubt as to what the complement of, say, yellow is, you have only to think of what hue would be produced by the two primaries *not* present. In this case, the two left out are blue and red. Mixed, they make violet. Violet is the complement of yellow and yellow the complement of violet. Red's complement is blue plus yellow, or green. Orange is a mixture of red and yellow; its complement is blue.

texture and pattern

The term *texture,* like many common words, has a simple meaning which has been complicated by an entanglement of connotations. Ordinarily, we think of texture as what appeals to the sense of touch, that is, the surface qualities of things, and in the strict sense there is nothing without texture. A coarse and nubby burlap has one sort of texture, a smooth sheet of window glass another, and the faint toothiness of kid leather still another. Yet, it is common to speak of some surfaces as being *more* textured than others. By that is meant, of course, that the character of a given texture is more prominent than that of another. For example, the roughness of the burlap is more attention-getting than the uninterrupted slickness of the glass. This usage is consistent with the original meaning of the word *texture,* which has the same root as the word ''textile'' and originally meant ''to be woven.'' It has since come to refer to the physical structure of any material and, by extension, to the surface properties of objects (Fig. 174).

Usually, we use the adjectives descriptive of the sense of touch to describe such visible properties as smoothness, roughness, grittiness, graininess, or velvetiness, but that is not always the case. One might also describe the appearance of a surface in terms of pure visibility, as being unbroken, irregular, dimpled, in sharp relief, and so forth. Still, it is simpler to convey precise meanings by taking recourse to the vocabulary of tactile sensation, for our eyes afford us the experience of touch even when we cannot reach the surface. Indeed, even art works that emphasize texture as an important feature of design rely upon the vicarious perception of feeling through sight, since handling paintings and sculptures is harmful to them. In the fine arts, visibility is all.

Collage, of course, was one of the first serious arts to maximize the concreteness of art works by the use of surface texture. Picasso's *La Suze* (Fig. 175) is made up of various kinds of papers cut into shapes which have been glued down on top of one another and enhanced by drawing. This not only gives the individual scraps a special, almost precious uniqueness but also emphasizes the way in which ordinary vernacular material is made artistic by becoming part of a design.

Some artists, like Spanish painter Tapies (b. 1923), have stressed the material nature of their works, building up strikingly textured relief surfaces from combinations of paint, sand, powdered marble, and other materials (see Fig. 176). On a far smaller scale and with infinite delicacy, Paul Klee portrayed a German singer of serious music by exercising the synesthetic appeal of his art. *Vocal Fabric of the Singer Rosa Silber* (Fig. 177) combines exposed linen canvas, fabric-textured *gesso,* and stains of watercolor *gouache* to suggest a lyrical sensibility. The lettering (which marries the singer's initials and the standard vowel sounds) is posed against all of this as if it were musical notation for a specific melody. Klee's textural experiment makes no attempt to copy the appearance of Rosa Silber; rather, it represents the auditory qualities of her singing style. Tapies simply substituted his textures for the textures of reality and made the surface the subject of the work, just as Mondrian replaced the world seen with his realm of severely stable stripes set against a void of white.

 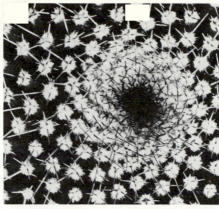 

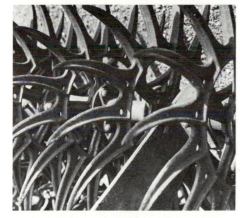 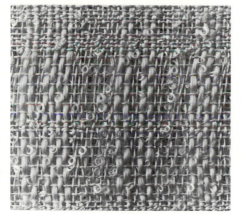 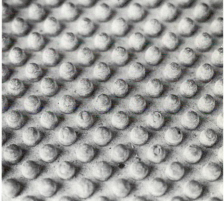

 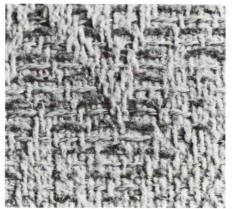

FIGURE 174 A sampling of textures

FIGURE 175 PABLO PICASSO. *La Suze.* 1912–13. Pasted on paper with charcoal, 25¾ × 19¾″. Washington University Gallery of Art, St. Louis

FIGURE 177 PAUL KLEE. *Vocal Fabric of the Singer Rosa Silber.* 1922. Gouache and gesso on canvas, 20¼ × 16⅜″. Museum of Modern Art, New York

FIGURE 176 ANTONI TAPIES. *Materia Organica Sobre Blanco.* 1974. Sand and mixed media on canvas, 51 × 64″. David Anderson Gallery, New York. Photo: Foto MAS, Barcelona.

Picasso has emphasized the textural variety of the paper by bringing different kinds together in a single work so that one notices such subtle differences as the contrast between slick, hot-pressed paper and the grainier cold-pressed fibers, between the pulpy, embossed wallpaper and the printed newsprint. Tapies, on the other hand, has practically sculptured his painting from sand-laden pigments, lending the dull richness of the work a massive, architectonic substantiality. Paul Klee, for whom the surface of a painting was always an important part of the image, does homage to a singer with a piece that has both the full-bodied vigor and the ineffable refinement of the woman herself.

Naïve amateur painters sometimes attempt to achieve realistic effects by using the pastiness of tube paints to imitate actual surface textures, in much the way cake decorators mimic rose blossoms with icing. Where they wish corn kernels to appear, they place similarly shaped blobs of yellow oil paint. Fur is simulated by their making thready strands of paint peak up from the canvas. The grain of driftwood is literally carved into thick gray paint. This approach is almost never successful in creating illusions, but it may have a charm of its own. Infrequently, professional artists will make somewhat similar use of the medium. Thus, van Gogh, in his *Bedroom at Arles* (Fig. 178), has mimicked the unpainted rough wood of his bedstead and the well-scrubbed floor through strokes of a bristle brush. The pillows and the bedclothes are wrinkled and the artist has carried these facts over into what he described as "flat surfaces, though coarsely brushed with heavy pigment," of this "interior without artifice."[1] The picture was done partly to decorate the house in anticipation of a visit from Paul Gauguin (1848–1903) and partly as an exercise, cramming as many complementary pairs in as possible. Both van Gogh and Gauguin were fond of this painting, but in their comments about it and the somewhat similar *Sunflowers* they emphasize the color and substance of the painted surfaces, not the imitative textural effects.

Van Gogh wrote of "plowing" his canvases as the peasants do their fields and of wishing to exaggerate things so that lines "are warped as in old wood."[2] He was conscious of the way in which such textural effects set pictures at a remove from nature. The textures of canvas, thick paint, brush marks, and palette knife strokes "label" the work as a creative act, to use Roy Behrens' words from Chapter 3.

Contemporary artist Robert Malone (b. 1933), in his painting *Model in the Doorway* (Fig. 179A), has combined objective, detached realism with the play of actual texture for its own sake by treating a studio wall as the wall itself rather than an illusion of a spattered wall. That is, the canvas is splashed, spattered, and plastered in exactly the way the real wall is supposed to be. It is not an imitation of that wall; it is an *equivalent* of the wall (Fig. 179B). The realism of the nude model, the stools, baseboard, and door jamb, then, turns the wall into a sort of visual pun or puzzle. Is the wall realistic or is it just reality? Is this a painting with collage elements or is it an extremely illusionistic depiction of a studio? An art work can, of course, be many things at once and this work is.

There is a whole category of still life painting which derives its aesthetic interest from a playful illusionism somewhat similar to Malone's, although in most respects *trompe l'oeil* (trick the eye) pictures are very different from *Model in the Doorway*. The whole point of such *trompe l'oeil* works as William Harnett's *Music and Good Luck* (Fig. 180) is to deceive the viewers by making them take for truth what is only a painted illusion. It is a mode that has rarely been

FIGURE 178
VINCENT VAN GOGH. *Bedroom at Arles.* 1888. Oil on canvas, 28 ½ × 36″. Art Institute of Chicago, Helen Birch Bartlett Memorial Collection

FIGURE 179A ROBERT MALONE. *Model in the Doorway.* 1981. Oil and acrylic with collage on canvas, 4'6" × 6'6". Private collection

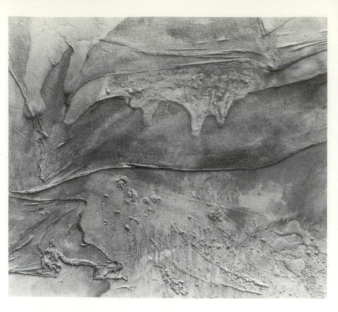

FIGURE 179B Detail of Fig. 178A

Van Gogh's painting of his bedroom was partly an experiment in combining flat zones of complementary colors, but the artist articulated those zones by texturing them according to the reality before his eyes. The bedclothing was wrinkled and the paint has therefore been manipulated so that it receives the same bends, folds, and crumplings. The bedstead was raw, unpainted wood, so van Gogh applied the pigment in such a way as to suggest a barely sanded rudeness. Malone's painting also uses actual texture to represent reality's texture. His nude model poses in a doorway opening through a paint-splashed, gauze- and plaster-patched wall before which two pristine stools stand on guard. In this case the wall depicted not only looks like the original wall, it even *feels* like the original. In fact, the only significant difference between the painting and reality, so far as gallery goers could tell, is that the former carries the textures on canvas and the latter on plasterboard or sheetrock panels.

FIGURE 180 WILLIAM HARNETT. *Music and Good Luck.* 1888. Oil on canvas, 43 × 30". Metropolitan Museum of Art, New York. Catherine Lorillard Wolfe Fund, 1963

practiced by professional artists and not once by a painter of major historical importance. Harnett (1848–1892) is probably the most skillful of the *trompe l'oeil* illusionists. Consider the date of his painting in the lower right corner. It is painted to look as if it had been carved into the wood. But is it to be thought of as Harnett's exact copy of a date he carved into the still life, or is it a date painted onto the picture in such a way that it is consistent with what the picture represents? This sort of mental game is not unlike the one Malone plays with his audience. Insofar as actual textures are concerned, however, the pictures are quite unalike, for the surface of the Harnett is extremely smooth regardless of the texture represented.

It is typical of extremely realistic renderings of texture that the actual texture of the picture is unbroken. The Harnett, van Eyck's *The Madonna of the Chancellor Rolin* (Fig. 181), and Holbein's *French Ambassadors* (Fig. 182) are all smooth, virtually without brush strokes, although each of them details many different textures with astonishing meticulousness. The reason for this is obvious, we realize. Smooth textures are the least obtrusive. They don't get in the way of illusion whereas more prominent textures do. Thus, the texture in Fig. 183A comes through with a

FIGURE 181 JAN VAN EYCK. *The Madonna of the Chancellor Rolin.* c. 1433–34. Oil on panel, 26 × 24³/₈″. The Louvre, Paris

FIGURE 182 HANS HOLBEIN. *The French Ambassadors.* 1533. Oil on panel, 81¼ × 82¼″. National Gallery, London

William Harnett's work is like camouflage insofar as it is deliberately deceptive. It would, however, be easy to detect the difference between the picture and an absolutely identical assemblage put together in a frame merely by changing your position slightly because the relationships among three-dimensional objects change when one's viewpoint changes whereas relationships among two-dimensional renderings do not. (That is, you cannot see the side of Harnett's matchbox by stepping to the left as you could if it were real.) That is one of the reasons that the most successful *trompe l'oeil* pictures depict things that are relatively flat.

For all of their precision, the van Eyck and Holbein could never persuade a viewer that they are pieces of reality in the way the Harnett might for at least a few moments. Nonetheless, each of the paintings in Figs. 180, 181, and 182 does represent the appearances of various textures in a convincing way. As is typical of such works, these are all smooth and fairly dark-complected.

certain authenticity when it is printed on smooth white paper but is muddled when imposed upon a fabric having a conspicuous weave (Fig. 183B).

Visual texture is communicated to us by light reflections reaching the eye. The ability of someone like Harnett to capture the fundamental difference between the soft highlights gleaming on ebony tuning pegs and sharp ones glinting from tautly strung wire is nothing short of miraculous. Van Eyck and Holbein discriminate exact distinctions among fur, velvet, and hair. The woods, the brass, and the fabrics in their paintings are distinguished by the most exactingly minute chiaroscuro.

Sometimes, however, the impression of exact fidelity to nature does not derive from counterfeit detail. Rembrandt's helmet of gold, for example, is rendered with brushwork quite unlike the jeweler's *chasing* on the helmet itself (Fig. 184). As Rene

Huyghe said: "At close quarters . . . illusion is dispelled; the pigment asserts itself to the exclusion of everything else. . . . We can gaze at it endlessly, as at the thousand shapes assumed by foaming waves or a crackling fire."[3] The marvelousness of this transformation of a texture seen up close into another when seen from a distance is surely captivating. It is, in effect, a special instance of the embedded figure (Fig. 185) except that in the painting the disparity of two ways of seeing the same pattern as contrasting textures is completely overwhelming. It is also much too complicated a phenomenon to brook analysis in this book. What is worth noting here is the relationship between texture and pattern.

Visual texture is entirely illusory, really nothing more than patterns of light and shade spread over a relatively unobtrusive surface. Compare Figs. 174, 179B, and 183A. Each of these textures can be de-

FIGURE 183A A texture

FIGURE 183B The same texture printed onto another, prominent texture

A smooth texture works best as the foundation of an illusory texture because it is least likely to interfere by interposing itself between the viewer and deception.

Things are not always as they seem. When one looks at the works closely, the treatment of brocades, jewelry, furs, and wood grain in the Harnett, van Eyck, and Holbein are what one might expect—tiny chiaroscuro versions of individual threads, gems, hairs, and so on—but the details of Rembrandt's *Golden Helmet* do not duplicate the appearance of a metalsmith's work magnified. Rembrandt's paint surface is full of twists and eddies which seem almost to spurt, seethe, and boil. But step back a couple of paces and these lava-like torrents collapse again into a resemblence of metal plates covered with gleaming foliate ornament. Unlike the earlier masters, Rembrandt did not copy precisely every small inflection of reality but instead left the viewer to reconstruct optically the texture of things seen.

FIGURE 184A REMBRANDT VAN RIJN. *The Golden Helmet.* 1646–47. Gemaldegalerie Staatliche Museen Preubischer Kulturbesitz, West Berlin. Photo: Jörg P. Anders.

FIGURE 184B Detail of Fig. 184A

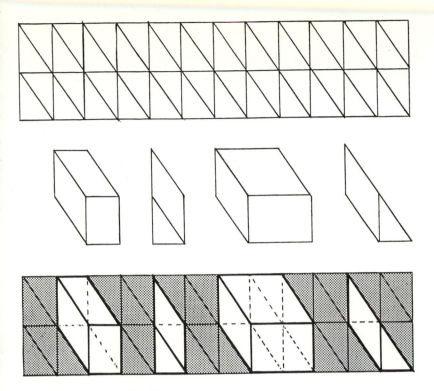

FIGURE 185 Embedded figure

scribed also as a pattern of light and dark. So can they all.

Patterns are, by nature, repetitious. Counterchange (Fig. 186) is one variety, texture another, and modes of pictorial composition as various as Coleman's (Fig. 187) or Reinhardt's (Fig. 188) could be considered special forms of patterning. In Fig. 189 the pattern of contiguous circles formed by the ends of sewer pipes overcomes most variations within the view. Cropping a segment, though (Fig. 189B), reveals some attractive textures concealed within the overall patterning.

In terms of purely visual phenomena, as opposed to actual touch sensations, there is scarcely any difference between texture and pattern. When a sheet of paper was folded into a pattern of diamonds and criss-crosses (Fig. 190), a texture was created that is not the same as the fiber structure of the paper. When Harnett discerned in the glossy varnished wood of the violin the repeated variations that gave it its singular complexion (Fig. 180), he perceived a natural patterning—wood grain—that could be duplicated through his craft.

There are times when the indefinite repetition of a single motif (as in Fig. 191A) is precisely what is called for. Even then the repetition is never truly endless; it terminates at the ends of a wall, against the ceiling, or down on the floor, and it is appreciated by us in juxtaposition with a contrast. Frequently, the most interesting use of pattern is in establishment of a system in which deviation occurs. Even a small change can be quite striking when it occurs in a monotonously regular context (Fig. 191B). Of course, that is only the most obvious kind of switch. An often cited example of a more imaginative variation on a pattern is *Flowing Phalanx* (Fig. 192) by Francis Celentano (b. 1928), in which black triangles and quadrangles produce a fluctuating dance of white triangles tied together by strings of white. An even more compelling impression of movement is evoked by Bridget Riley (b. 1915) with *Current* (Fig. 193), an example of Op Art.

Establishment of a strict pattern as a springboard for variation is also the basis of Jasper Johns's *Numbers in Color* (Colorplate 1), Agnes Martin's grids (Fig. 194), and Vasarely's *Vega* (Fig. 195). François Robert's design for Dunbar Corporation (Fig. 196) is based on the same notion. In an extreme extension a potter and sculptor named Paul Dresang (b. 1948) has disrupted a pattern of a more general sort. By the simple but very ingenious ploy of squeezing flat vessels that he has formed from clay in the usual manner, Dresang has created a series of ceramic reliefs (Fig. 197) which not only are appealing as objects, but also make a whimsical comment on the nature of clay bodies. The texture of a clay body formed on a

FIGURE 186 Counterchange

FIGURE 187 FLOYD COLEMAN. *Neo-African #12.* 1977. Mixed media on paper, 9½ × 9½″. Private collection

FIGURE 188 AD REINHARDT. *Abstract Painting, Blue.* 1952. Oil on canvas, 75 × 28″. Museum of Art, Carnegie Institute, Pittsburgh

Patterns, of which all of these images may be considered some kind of example, are similar to texture and, indeed, overlap with it. In fact, patterns impose their character upon the patterns of the textures which constitute their small elements. That is particularly obvious in the photograph of the sewer pipes, where the overall pattern completely dominates everything else, and in the sheet of folded paper, where the indentations become the principal texture.

FIGURE 189A JOHN ADKINS RICHARDSON. *Charlene of the Rings.* 1957. Photograph, 7½ × 8¼". Private collection

FIGURE 189B Detail of Fig. 189A

FIGURE 190 Paper sculpture

a

b

FIGURE 191 Wallpaper pattern A: Indefinite repetition of motif; B: Unexpected deviation from established pattern

Pattern is based on the idea of elements repeated over and over again, but it is frequently the case that more interesting, even arresting, results can be achieved by deviating from repetition to suggest rhythmic movements.

FIGURE 192 FRANCIS CELENTANO. *Flowering Phalanx.* 1956. Synthetic polymer paint on canvas, 34 1/8 × 46 1/8 ". Museum of Modern Art, New York (Larry Aldrich Foundation Fund)

FIGURE 193 BRIDGET RILEY. *Current.* 1964. Synthetic polymer paint on board, 4′10 3/8 ″ × 4′10 7/8 ″. Museum of Modern Art, New York

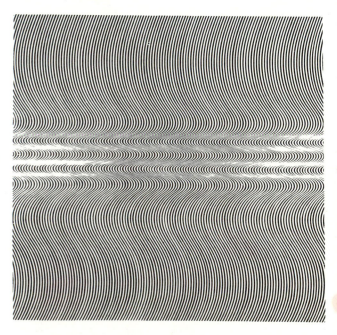

FIGURE 194 AGNES MARTIN. *The Tree.* 1964. Oil and pencil on canvas, 6 × 6′. Museum of Modern Art, New York. Larry Aldrich Foundation Fund

FIGURE 195
VICTOR VASARELY. *Vega.* 1957. 77 × 51″. Collection of the artist

FIGURE 196
FRANCOIS ROBERT. Booklet cover. Dunbar Corporation

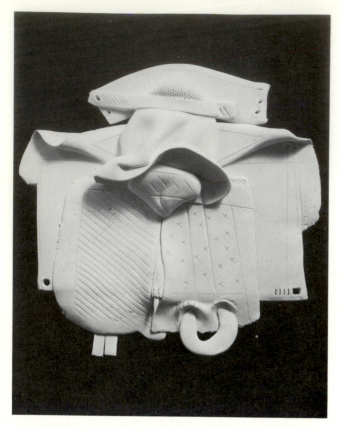

FIGURE 197 PAUL DRESANG. *Untitled.* 1981. Unglazed porcelain, 9 × 10″. Private collection

FIGURE 198 DANIEL ANDERSON. *Bottle.* 1982. Raku body and low fire glaze, height 13″. Private collection

Normally we think of pottery, at least the pottery made on a wheel, as being round and hollow. That is the expected pattern. But Paul Dresang has gone contrary to expectation and flattened his pots into relief sculpture before drying and firing them.

potter's wheel and fired is one we associate with hollow, free-standing utensils such as pitchers, bowls, and flowerpots (Fig. 198). Dresang's reliefs run counter to the established pattern.

Painter William Bailey (b. 1930) has exaggerated the expected patterns of pottery—cylindrical form, earth colors, dull, grainy textures—by doing still lifes which emphasize the commonality of clay surfaces in softly graduated light and shade (Fig. 199). All still life painters attempt to find within their subject matter similarities of shape and repeated notes of color which will provide harmonious patterns on which to base compositions. Harnett's horseshoes and hinges repeat curves of the violin; even a work as prolix in detail as Jan van Huysum's *Flowers and Fruit* (Fig. 200) attends to compositional niceties through the pattern of repeated curves in the flower stalks and in the sculptured vase containing them.

Finally, to close off this discussion, we come to a work which outreaches all of the others in its oblique approach to the representation of textures. *Variations on a Millstone* (Colorplate 21) by Stephen Posen (b. 1939) is a deliberate exercise in balancing illusion against reality. Study the painting. Its complex play of textures and contradictory spaces is quite disconcerting. On the one hand it is photographically realistic and, on the other, impossibly queer. Here's how it was done: Posen began with a black and white photograph of great resolution and then enlarged it enormously, until it was just over seven feet tall. He then created a collage by stapling an arrangement of colored fabric strips to the photograph. Using this collage as a model, he painted a picture of exactly the same size as the enlarged photograph, and it is this painting that is reproduced as Colorplate 21. Beginning with photographically objective visual textures, Posen added physically tactile elements, and ultimately reproduced the whole by hand. This quirky illusionism takes on the complexities usually encountered only in philosophical discussions of things like epistemology.

FIGURE 199 WILLIAM BAILEY. *Large Umbrian Still-Life.* 1973. Oil on canvas, St. Louis Art Museum.

William Bailey heightens the customary composure of still life by eliminating from his pottery still life everything but simple geometric objects of similar textural appeal. This is not altogether new—a prototype can be found among the works of Spanish artist Francisco Zurbaran (1598–1664), for instance—but it contrasts startlingly with traditional still life painting. Flower painter Jan van Huysum (1682–1749) is at the extreme opposite from Bailey. His paintings are incredibly detailed conversation pieces meant to startle audiences with the sheer density of charming, nearly overlooked details, but the profusion of visual material is contained within a design made up of repeated curves and colors and textures. In Bailey's picture the pattern is ascendent over everything else; in Huysum's the pattern is subsumed beneath a multitude of particulars.

FIGURE 200A JAN VAN HUYSUM. *Flowers and Fruit.* c. 1736 Oil on canvas, 52⅝ × 35⅞″. National Gallery, London

FIGURE 200B Detail of Fig. 200A

EXERCISE 7–1: Textural Collage

Materials: White drawing paper
Colored construction paper
Rubber cement
White all purpose glue (e.g., ''Elmer's Glue-All'')
Variously textured cloth and paper
Picture magazines and newspapers

Tools: Single-edge razor blade or frisket knife
Scissors
Graphite sticks
Cutting surface

Using illustrations from magazines, newspapers, and so forth, as well as frottage rubbings and bits of cloth and decorative papers, create a composition in which textural variation is emphasized. Fig. 201 provides some student works of the type.

FIGURE 201 Student textural collages

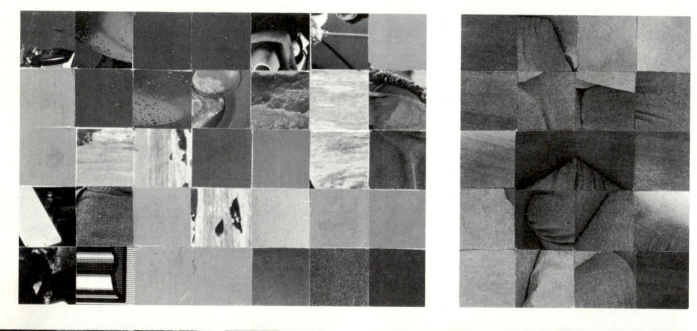

EXERCISE 7-2: Found-Object Collage

Materials: *Heavy sheet of cardboard, plywood, etc.*
Variously textured pieces of rubbish
White all purpose glue (e.g., ''Elmer's Glue-All'')

Tools: *Varied, depending upon materials used*

Fig. 202 is a photograph of some collages made up of various bits of rubbish combined with nature's dross. The materials have been glued and tacked to boards in essentially the same way in which the things in Exercise 7–1 were assembled. Some of the combinations of materials, such as the nylon and paper, are quite subtle in effect while others are highly charged. Whether they are bold or ethereal, though, the juxtapositioning of materials should emphasize textural particularities that would otherwise be less evident to the viewer.

FIGURE 202 Student collages from found objects

EXERCISE 7-3: Rendering Textures

Materials: *Appropriate size illustration or canvas board*
Designer colors

Tools: *Brushes*
Colored pencils and various pens, as required

Attempt to reproduce the appearance of the collage you did for Exercise 7–2 as precisely as possible, using designer colors in combination with whatever other drawing and painting media you wish. See Fig. 203 for examples of the exercise as carried out by others.

In undertaking this exercise it is very important to be sure that your model—the collage—is illuminated by a constant light source so that neither the direction nor the intensity of the light varies appreci-

FIGURE 203 Student renderings of collages from found objects

ably. That is nearly always a consideration when one undertakes an exact recording of textural effects and it is one reason that artists and designers usually try to secure studios having a northern exposure. Sunlight entering a room through windows facing north is more constant throughout the day than is sunlight falling directly through the glass on the south.

Artificial lighting may and probably will be used for this exercise, but sunlight is much preferred since things seen under incandescent or fluorescent light are significantly altered in hue, intensity, and value. The whiter the light, the better.

EXERCISE 7-4: Sandcasts

Materials: *Clean, fine sand*
Stones and other small objects
Plaster of Paris
Petroleum jelly

Tools: *Plastic bucket*
Large commercial baking pan or similar container
Random sticks and ''stuff''

The main portion of the unusual fireplace in Fig. 204A, the firepot, resembles a large stoneware ceramic; actually, it is formed from a concrete-like refractory substance that is much stronger than fired clay. The tiles ornamenting the wall behind the firepot look like segments from a beach and feel like hard, dry slabs of sand, and in fact the surface is made of sand. The detail in Fig. 204B gives you a somewhat better impression of the character of these panels.

Sandcast items need not be as simple as these are. They can be made to stand out in much higher relief and can contain a far greater variety of materials.

Sandcasting is also quite easy. You need a container to serve as a mold for the sand. It can be nothing more than a hole in the ground with some boards for walls and the sand for a floor. The tiles in Fig. 204 were cast in large commercial baking pans. Put slightly moistened sand into the bottom of the container and press stones, shells, and other objects down into the sand. Pour Plaster of Paris of casting consistency into the container and allow it to set for a few hours. When the plaster is lifted from the mold, it will have hardened around the portion of each stone that was protruding from the sand and will therefore hold fast the stones. Since grains of sand are nothing but tiny stones, the uppermost layer of sand will be picked up by the plaster in the same way. The back of the plaster slab is smooth and white but the front has the color, appearance, and texture of a

FIGURE 204A Fireplace and sandcast tiles. Ravine house. Edwardsville, Illinois

stone-littered beach. Since only a tiny amount of sand has been picked up, the remainder can be smoothed out and used again. A few inches of sand will last for a long while used this way.

Bear in mind that the casting will be the reverse of the mold. If you poke a hole into the sand with your finger, the casting will come out as a finger-shaped protuberance. The part of the stone or other object that will show up in your casting is the side you *bury* in the sand and not the side you see. What you put down on the right will turn up on the left of the tile. Also, be sure to make the tiles fairly thick—at least 1 1/2 inches—so they will not be too fragile. Remember too that it is better to let the plaster dry too long than to remove it too soon. Finally, it will be easier to free the casting from the mold if you put a very light layer of grease on its walls.

Plaster of Paris is sold as a flour-like powder. It is easy to mix up but it is rather messy and certain

FIGURE 204B Sandcast tile from Fig. 204A

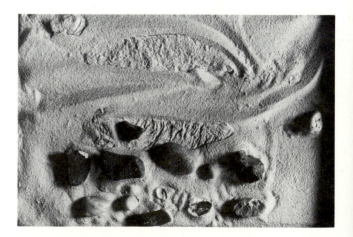

procedures must be followed. Extra plaster is wasted since it dries very fast and cannot be reused. Fill a cheap plastic bucket with enough water to fill the space you will using. (You can determine this by first filling the empty mold with water as deep as the thickness you would like the casting to be, then pouring the water into the bucket. This will be a little more water than each tile will require.) Dump powdered plaster into the water by the handful. The plaster has little effect on the water level, because plaster absorbs much more water than it displaces. When you have filled the water with plaster just up to the surface, so that little islands are standing above the liquid in a few spots, mix them together. It's best to use your hands, squeezing the plaster into the water with your fingers, generally moving from the bottom of the bucket upward. You *can* stir the plaster instead, but you must use only a round stirrer such as a dowel rod between 1/2 and 1 inch thick. Flat sticks create turbulence in their wakes and therefore make bubbles, which are disastrous in plaster.

When the plaster is smooth and creamy, it is ready to pour. Easy, steady pouring will not disturb the sand and will give you what you formed in the mold. However, sometimes a splashy, sudden pour produces interesting configurations and, with a little practice, one can learn to control such things just as Pollock and Sam Francis did their drippy paint.

Don't dump leftover plaster in a sink! It will solidify there, even in water, and will completely clog the pipes, necessitating dismantling and possibly replacing the plumbing. Dump the plaster onto an open newspaper several sheets thick. Scrape out as much as possible from the bucket. Then rinse out the bucket, using lots of water to dilute the remaining plaster to the point where it barely exists. Fold the newspapers up and toss them into the trash. The other household chores connected with sandcasting are routine and obvious.

Obviously, the exercise is to do some sandcasts.

If you wished to make particularly decorative patio or walkway tiles, you could use this same technique, substituting ready-mix concrete for plaster.

EXERCISE 7-5: Programming Patterns —Variable Composition

Materials: *White drawing paper*
Picture magazines
Rubber cement
Tools: *Single-edge razor blade or frisket knife*
T-square and triangles
Ruler with metal edge
Cutting surface

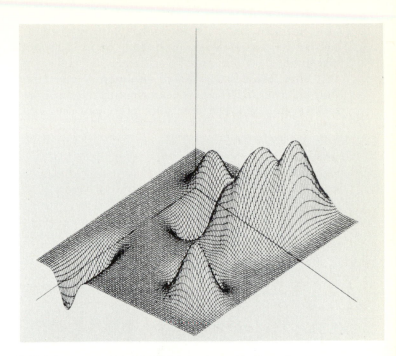

FIGURE 205 CHARLES CSURI. *Peaks and Valleys.* Computer drawing of a surface based on four boundary curves.

A great deal of design work today is carried out by computers. These electronic brains can be instructed to direct pens set into mechanical arms which will draw blueprints and even do perspective renderings of buildings and other objects much more rapidly and with far greater accuracy than could the most skilled and tireless draftsmen. The computers work with instructions that are very simple in their specific elements—such as A = Yes, AA = No or HOLE IN CARD = 1, NO HOLE IN CARD = 0—but are extremely complicated in their number and interconnections. The pattern of instructions is called a *program.* Most design systems can be programmed for computers (see Fig. 205) and, while we do not propose actually to do that here, we can approximate the more obvious traits of such programming by means of some simple exercises.

Devise a system of sixteen squares, each square 1 1/2 inches on a side, arranged in a large square 6 inches on a side. The design of each of the small squares must be different from all of the others and yet must resemble all of them, too. The idea is to create a system of modules such that a pleasing combination will result when they are arranged in a 6-inch square *no matter how a given square is turned or what position it occupies.* The resultant composition should be like a game with seemingly endless, interconnected, exciting solutions.

In order to make the problem less tedious, we are permitting you to use two blank squares of a single flat color. You may create as many as twenty units to draw upon for your variable composition, but no more than sixteen units may be used at any given time.

Fig. 206A is a series of variations created by shifting around of the same sixteen units. None of the units is precisely like any other, although they are very similar. Fig. 206B is a variable composition using components that might easily be combined with those of Fig. 206A. In Fig. 206C we see some of the more elaborate kinds of motifs students have come up with. Fig. 206D is a serious painting using four panels in a similar fashion.

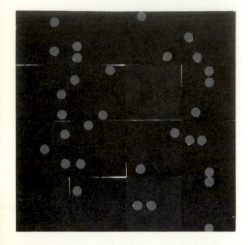
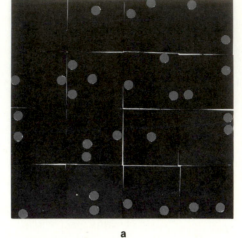
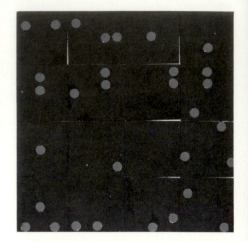

a

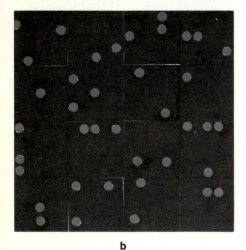

b

FIGURE 206 Variable compositions. A: Several variations; B: With additional components; C: Student motifs for exercise;

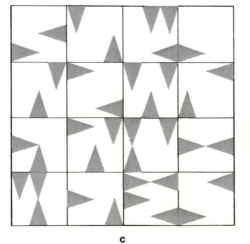

c

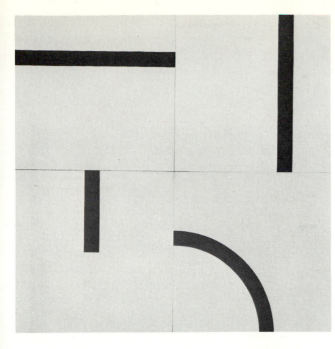

D: MICHAEL SMITH. *Painting.* 1970. Oil on canvas 4′ x 4′. Private collection

EXERCISE 7-6: Programming Patterns —Modulated Variations

Materials: White drawing paper
Tools: T-square
Triangles
Ruler with metal edge
French curves

In Fig. 207A you can see a module, and such modules are arranged into a pattern in Fig. 207B. In Fig. 207C the same module has been pivoted in the second register so that a new pattern emerges. In Fig. 207D every other module has been shifted. Fig. 207E is a more involved mutation.

Create a module having a similar capacity for producing a multitude of patterns merely by being

FIGURE 207 A "programmed" pattern. A: Module; B: Pattern #1; C: Pattern #2; D: Pattern #3; E: Pattern #4; F: Pattern #5.

a

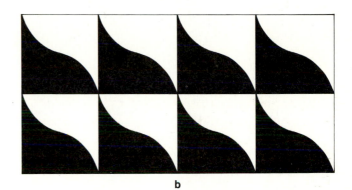

b

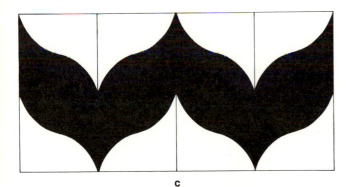

c

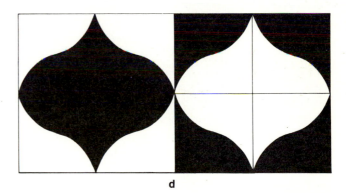

d

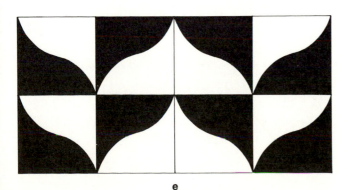

e

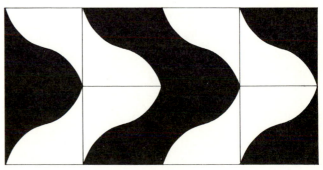

f

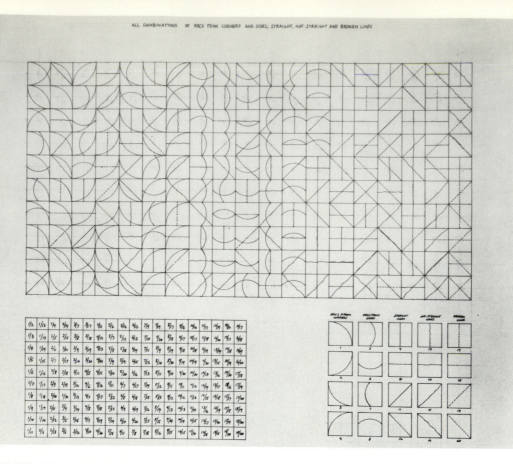

FIGURE 208 SOL LEWITT. *All Combinations of Arcs from Corners and Sides; Straight, Not-Straight, and Broken Lines.* 1976. Ink and pencil on paper, 22³/4 x 26⁵/8″. New Britain Museum of Fine Arts, New Britain, Conn.

rotated on its axis. You will discover that when all four axes interlock into overall patterns the rhythmic alteration of one kind of pattern into another is "programmed" into the grid. Try turning the second module a quarter turn clockwise and inverting the third, leaving the fourth alone, and repeating the previous treatment of second and third with the fifth and sixth. Continue on in this fashion; it is the procedure that produced Fig. 207F.

Experiment with other modifications of position, recording them in some way as you proceed. What are the controlling conditions? Is there a pre-dictable consequence of certain procedures? What governs it?

Can you see how this problem relates to the kind of things computers do? The work of Sol LeWitt (b. 1928) is a thoroughgoing investigation, which he says is really mystical rather than rational, into the nature of modular change or permutation. Fig. 208 is a plan for a drawing done on a wall for the Venice Biennale in 1976. As you can see, it is an elaborate design born of twenty units and a program for combining them. LeWitt is a Conceptual artist whose media are as varied as drawings, murals, string designs, sculptures, and performances.

notes for chapter 7

[1]Vincent van Gogh in a letter to Paul Gauguin (Arles, October 1888) quoted in John Rewald, *Post-Impressionism from van Gogh to Gauguin* (New York: Museum of Modern Art, 1956), p. 242.

[2]Vincent van Gogh, *The Complete Letters of Vincent van Gogh* (Greenwich, Conn.: New York Graphic Society, 1959), vol. 3, pp. 226 and 217 respectively.

[3]Rene Huyghe, *Ideas and Images in World Art* (New York: Harry N. Abrams, Inc., 1959), p. 199.

space

Posen's *Variations on a Millstone* (Colorplate 21) is perplexing because of its spatial characteristics. The black and white background suggests three dimensions while the pieces of colored cloth seem to be laid out upon a flat surface. A perfectly plausible explanation of the space, one that has nothing to do with artistry—but that occurs to practically nobody—is that the ribbons are attached to a sheet of glass. Artists have, in fact, often been intrigued by like ambiguities produced by transparency. Richard Estes (b. 1936) specializes in sharp-focus copies of color photographs featuring shop windows and he is fascinated by reflections. In his *Ansonia* (Fig. 209) the right-hand portion of the painting mirrors the left. We see not only the reflection but also the plants beyond the glass and even the backside of the window to which various decals have been applied.

Estes's picture is not disturbing in the way Posen's is because he has related the complex overlappings to a common visual experience by giving them a familiar context and showing us the basis of the mirror image so that we can compare reality with its illusion. The "real" image in *Ansonia* is, of course, actually just as much an illusion as the reflection on the window. Estes painted both of them. Not only that, he took them from yet another illusion—a photograph. The painting itself is entirely flat, just as paintings by Mondrian (Fig. 210) and Picasso (Colorplate 14) are.

Usually, we do not experience any confusion over the spatial content of a painting or drawing, because artists tend to select one of two alternatives when they create pictures. They treat the surface of the work as if it were a window through which the world is seen or they treat the surface as a screen upon which forms are placed. Psychologically, it is the same with paintings as with real screens. If you focus on the world beyond the screen, you are not much aware of the screen itself (Fig. 211A); if you focus on the screen (Fig. 211B), the world beyond has little prominence. Estes does the first, Mondrian the second, whereas Posen has gone out of his way to combine the two alternatives. The matter is not, however, a simple one, for even the Mondrian suggests a limited kind of space. It is severely limited in depth, perhaps, but its articulation of the two-dimensional plane of the canvas involves something more than just dividing up an area into pleasing proportions.

Fig. 212A contains typed lines of Xs which establish a certain kind of spatiality for the rectangle containing them. Fig. 212B is the same rectangle and it originally contained identical lines of type. When the ink lines were added, connecting groups of Xs at different levels, the amount of white space in the rectangle was in fact *reduced*. Obviously, there is literally less white space in B than there is in A. But it doesn't look that way, does it?

Even very simple light and shade relationships can radically change one's interpretation of picture space. Fig. 213A is identical to 213B; we have merely inverted the photograph for B. One's unconscious presumption that light falls upon things from above rather than from below produces the misinterpretation. If you wish to be sure that a picture with an unusual light source will be interpreted correctly, you must be quite obvious about it and very consistent, as

FIGURE 209 RICHARD ESTES. *Ansonia.* 1977. Oil on canvas, 48 × 60″. Whitney Museum of Modern Art, New York

FIGURE 210 PIET MONDRIAN. *Composition in Black, White and Red.* 1936. Oil on canvas, 40¼ × 41″. The Museum of Modern Art, New York. Gift of the Advisory Committee

FIGURE 211A Focusing through a screen onto the objects beyond pretty much negates our consciousness of the screen itself.

FIGURE 211B Focusing on the screen makes the world beyond fade into incidental background.

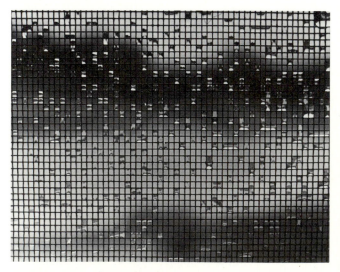

FIGURE 211C Closer focus on screen makes landscape still more obscure.

Correggio (1494–1534) was in his famous *Adoration of the Shepherds* (Fig. 214). This painting, popularly known as *Holy Night,* contains an incandescent Saviour whose radiance has been copied in innumerable holiday tableaux all over Europe and America during the Christmas season.

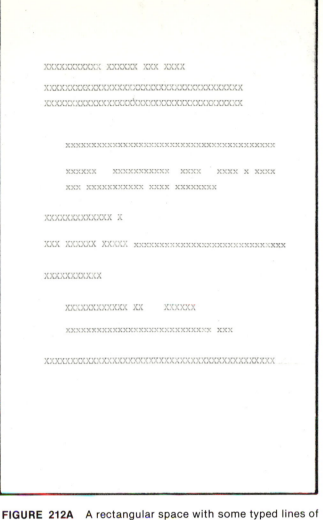

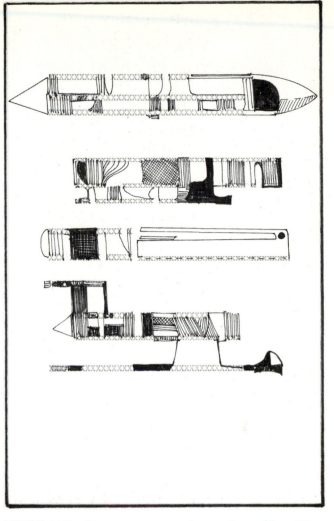

FIGURE 212A A rectangular space with some typed lines of Xs in it.

FIGURE 212B The same space and same lines of typed Xs, but with the lines connected up with ink marks.

Both of the pictures below are of the same decorative sculpture by American architect Louis Sullivan (1856–1924). Fig. 213B seems like a negative mold or cast of Fig. 213A only because it has been inverted.

FIGURE 213A LOUIS SULLIVAN. Ornamental panel from the Schiller Building. 1892. Terra cotta, 26½ × 26½". Collection Southern Illinois University at Edwardsville.

FIGURE 213B Figure 213A inverted.

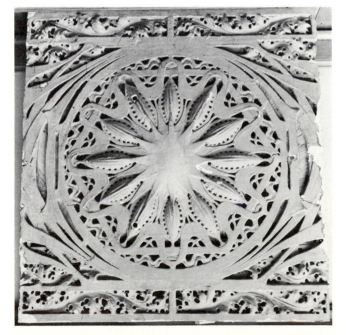

FIGURE 214 CORREGGIO. *Adoration of the Shepherds (Holy Night)*. 1522. Oil on panel, 8'5" × 6'2". Gemaldegalerie, Dresden (Alinari-Scala)

The relationship between space and volume is a very special one. Look at the sets of forms in Fig. 215. There is a Y, then a hexagon, then the two combined. The combination also resembles a cube, of course. With one segment darkened, the cube is even more convincing. With one side black and one gray, the effect is quite compelling. But whether the three faces of the box seem to be the top and sides of an exterior, or seem instead to form the interior walls and ceiling of a box seen into from below, is just a matter of whim. When we see such boxes fitted together into a honeycomb pattern we have no difficulty conceiving of them as a stack seen from above or, conversely, as a cluster hanging over us. Switching the light and dark surfaces around fails to destroy the three-dimensional effect, although it turns what was ambiguity to confusion. The relevance of this factor to representations of space in drawing and painting is made clearer in Fig. 216, where identical background planes have become (A) an interior and (B) an exterior by virtue of the figures seen against them. These characters from a Western melodrama help us to establish a viewpoint, and there is nothing more important in comprehending optical space. Even relative size is not as important.

The *most* fundamental spatial relationship, optically, is overlapping. One of the first things baby animals learn is that what is ahead of something else usually blocks it out. It is possible to imagine that in

FIGURE 215 A: Y; B: Hexagon; C: Y in a hexagon; D: Cube; E: Cube; F: Pattern of cubes; G: Alternative pattern of cubes

In painting and drawing, whenever a light meets a dark, an accent is obtained, just as a contrast in tone creates an accent in a musical sequence. When the accent is formed by the junction of two areas, one uniformly dark relative to another which is uniformly lighter, an apparent change of plane occurs. For this reason, the cubes in D and E are more convincing than the one in C. Whether the form such cubes appears to take is in the nature of protrusions or indentations, however, depends upon many other things. The pattern in F is ambiguous; we can make the forms go in or out. The suggestion of a third dimension does not depend upon the presumption of a light source as much as one might suppose. Even when we utterly disrupt the obvious regimen of F with some other program for the white-black-gray pattern (G), the impression of a relief surface remains, although its nature is then muddled rather than just ambiguous.

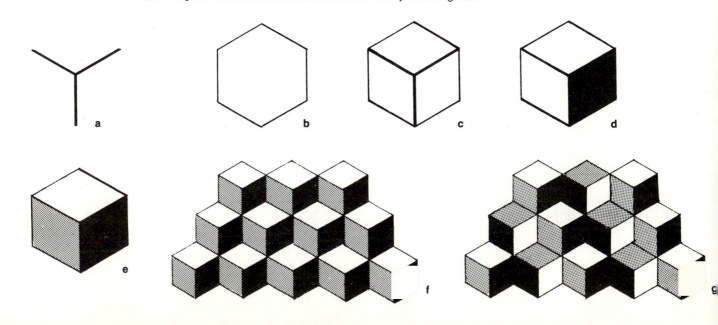

a

b

FIGURE 216 The backgrounds are identical in the two drawings but we see them differently because in A we identify with the figure staring over us and think of the black area as a ceiling, whereas in B we are looking down onto the figures from and, therefore, interpret the black area as a rooftop or cast shadow.

Fig. 217 the P is a small, inverted L shape ahead of O and that Q is another L in front of both P and O. However, the normal, immediate reaction is that O overlaps rectangle P, which is in turn overlapping rectangle Q. Even so simple a diagram as Fig. 217 requires viewers to assume that it represents geometric forms being observed from a specific viewpoint. It does not matter whether the interpretation is the conventional one or an eccentric one; either assumes a viewpoint. If 217 *were* made up of inverted Ls with Q nearest the viewer, it would look like Fig. 218A if seen from the left, whereas the same viewpoint would reveal something quite different (Fig. 218B) if Fig. 217 were what one usually thinks it is supposed to be.

FIGURE 217 Overlapping planes

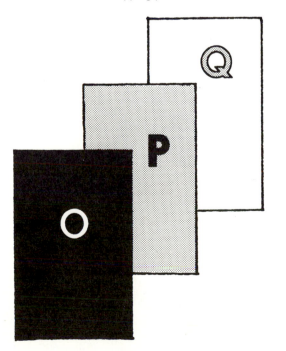

FIGURE 218A Fig. 217 as it might look from a different angle, given an eccentric reconstruction.

FIGURE 218B Fig. 217 as it would look from a different angle, given the conventional interpretation as correct.

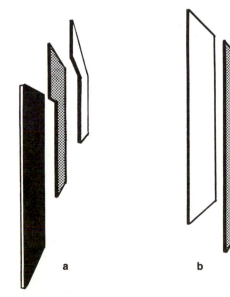

a

b

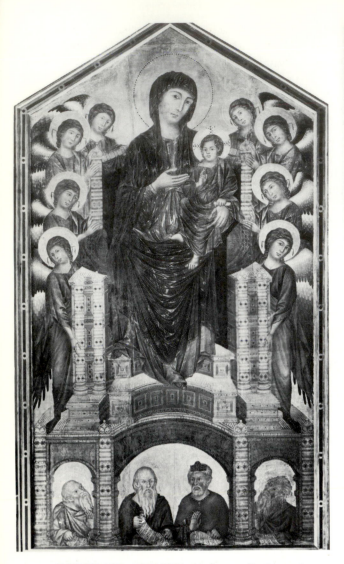

FIGURE 219 CIMABUE. *Madonna Enthroned.* c. 1280–90. Tempera on panel, 12′6″ × 7′4″. Uffizi Gallery, Florence

FIGURE 220 GIOTTO. *Madonna Enthroned.* c. 1310. Tempera on panel, 10′8″·× 6′8″. Uffizi Gallery, Florence

These pictures appear similar at first because they both use size to signify importance and neither is in perspective, a system unknown during the lifetimes of the painters. A basic difference between the two is the clear indication of a specific viewpoint.

One's first impression of the Cimabue is perhaps that we are meant to be looking at it as if from dead center, but that impression is false. If it were correct, we would not be able to see the inside of both sides of all three arches at the bottom. With respect to relative height, we can assume we are on eye level with any of the angels, with the Madonna, or with the prophets. The only consistent pattern is of looking down into the elements of the throne from the same angle no matter how far down or up the elements are.

Giotto has done numerous things to make his picture seem like a segment of reality. We are able to see just a trifle more of the left side of the bottom step on the throne, which means we must be standing slightly to the left. As for relative height, we can see the tops of the throne arms, which means our eyes must be above them. Also we can see the underside of the canopy, which means our eyes must be below that. A number of other advances make the Giotto seem more realistic. The light comes from a definite light source on the right, whereas in Cimabue light and dark are treated the way they are in Fig. 144. Giotto has massed the figures toward the bottom of the picture, suggesting gravity at work, and he has played up the inside/outside surfaces of the Virgin's cowl. These things, among others, make for a feeling of being presented with an extension of one's own space.

In order to see how viewpoint relates to the development of illusionistic rendering, it is instructive to compare and contrast a pair of late medieval paintings now displayed in the same room in the Uffizi Gallery in Florence, Italy. Both of them imply relative size in the way the Middle Ages frequently did, to indicate the relative importance of the people depicted, and neither painting is what we would call "realistic." But in the relationship between the two works we can see the emergence of a new way of depicting the world.

About thirty years after Cimabue (c. 1240–1302) painted his *Madonna Enthroned* (Fig. 219), his pupil Giotto (c. 1266/67–1337) did a rather similar one (Fig. 220). The latter is vastly more advanced in the direction of Renassiance realism. This is not because Cimabue wasn't striving. His painting reveals him fumbling after an illusionistic space in the lower part of the work. The concave steps below Mary's feet are an indication of his attempt. He tried to carry the effect on down into the arch above the anonymous prophets' heads, but the marriage of arches and concavity gave birth to ambiguity. Is the arch like the ones that flank it, or is the central arch an indentation that just happens to resemble the flanking arches? No one can say.

Giotto didn't know perspective drawing any more then Cimabue did—it wasn't invented until eighty years after his death—but he did establish a pretty clear vantage point. The viewer (who is the same person as the artist so far as the artist is concerned) is conceived of as standing slightly to the left of center and being approximately as tall as the two angels standing on either side of Mary's throne. This explicitness is very important. Cimabue gives you no good clues about your position. Giotto does. His illusion is incomplete and unconvincing to us because we are surrounded by images like the one in Fig. 209. However, without Giotto or someone like him, photography might not have come to be, for Giotto was moving in the direction of perspective drawing and there is scarcely any difference between perspective and photography so far as the outlines of things are concerned. Albrecht Dürer once did a woodcut showing, in one quick step, how perspective works (Fig. 221). In terms of contours photography does essentially the same thing with a convenient mechanical device instead of a cumbersome setup of plumb lines, hinged frames, and cross hairs.

Perspective drawing is like photography because both are products of the same mathematics. Neither is really like human vision, though. In the

These men are doing a drawing of a lute as it would look to us If we were standing with our eye at the spot on the wall marked by the hook. One man attaches the string (which represents a light ray) to the point on the lute to be noted. The other man drops a plumb down to where the string passes through the frame (representing a picture surface), then loosens the string and swings the drawing board around to mark the dot. Such a procedure will give an accurate perspective drawing of the lute once the dots are all connected.

FIGURE 221 ALBRECHT DURER. *Demonstration of Perspective, Draftsman Drawing a Lute.* From the artist's treatise on geometry. 1525. Woodcut, 5¼ × 7⅓". Kupferstichkabinett, West Berlin

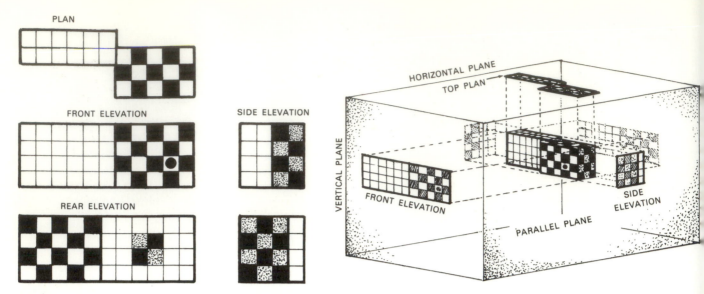

FIGURE 222 Plan and orthographic projection

FIGURE 223 The relationship of an orthographic projection to a three-dimensional object it represents. Each point on the object is recorded onto a plane in scale without regard to appearance.

An offset stack of cubes having sides of white, black, and gray clearly demonstrates the relationships among different projections.

first place, most of us see binocularly, that is, with two eyes, and perspective and photography are usually monocular. An entire photograph is in focus from top to bottom and edge to edge, at least within what is called its *depth of field.* Usually, in perspective drawings everything within the picture is in the same degree of clarity. Our eyes don't work that way. They pick out a tiny portion of the whole scene and focus on it. The eyes are constantly moving and we combine hundreds of different focused views into a mental image that seems similarly clear because of the rapidity of thought. A photograph is like a generalized statement of what one moving eye would see from a fixed position. Scientific perspective is an outline diagram of the objects as they would be seen that way, too. It is a specialized form of projective geometry, and it is but one of many ways to project objects in space onto a two-dimensional surface.

Consider Fig. 222. It couples a plan, at top, with *orthographic projections* in the form of *elevations. Ortho* means, in this usage, "at right angles" and that is the relationship of the diagrams to the three-dimensional objects they represent (see Fig. 223). All of the different kinds of projections we are going to be discussing can be derived from orthographic projections. Even engineers speak casually of elevations as "views," but they are not views in any real sense. If the side elevation in Fig. 222 were a view of the cubes, it would look something like Fig. 224. An elevation gives you a diagram of measurements; if something on the side is one foot tall, it is drawn to the same scale as everything else that is to be one foot,

whether it is on the nearest face, 6 inches back, or 200 feet away. A "plan drawing" is the most common kind of orthographic projection—a sort of topside elevation, usually with interior divisions indicated. Plan drawings are made to be worked from rather than looked at, but even carpenters, machinists, and masons find it helpful to have pictures of the things they are to construct. For their purposes a method of drawing pictures that offers a high degree of accuracy and clarity is required. There are three broad techniques of technical illustration that provide this kind of clarity. All are forms of engineering drawing and none requires the slightest talent for freehand drawing.

Oblique projections (Fig. 225) simply combine the plan with the front and side elevations from the original diagram. They give a fairly good approximation of how the object will look when completed and also provide measurements in scale. There are two kinds: *cavalier projection* and *cabinet projection.* They are not particularly popular, however, because the first looks "wrong" and the second requires that side and top measurements undergo at least one translation, into halves, even when the cabinet projection is the same size as the object itself.

Planometric projections (Fig. 226) take the measurements of the elevations and the plan and rotate them at an angle. If the vertical scale is reduced, a planometric projection is then referred to as being in *military perspective.*

Axonometric projections (Fig. 227) are the most widely used forms of engineering drawings and are of

154   ELEMENTS

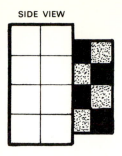

SIDE VIEW

FIGURE 224 A side *view* of the stack of cubes diagrammed in Fig. 222. Notice the small size of the checkered extension in contrast with the measured scale of these same things in the side *elevation.*

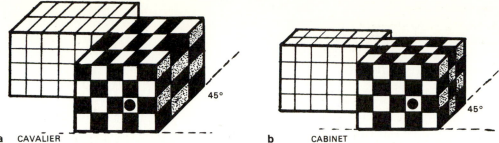

a   CAVALIER   45°   b   CABINET   45°

FIGURE 225 Oblique projections of the top, front, and left side. A: Cavalier projection: All measurements are to the same scale as the plan. B: Cabinet projection: Frontal measurements duplicate the plan but depth is in a ratio of 1/2 the scale of the plan to give an impression of greater realism.

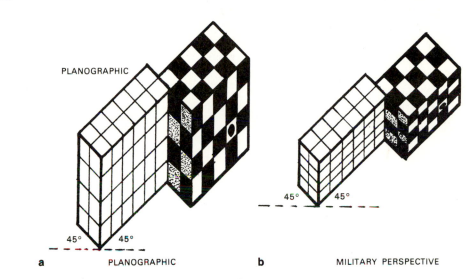

PLANOGRAPHIC

45°   45°   45°   45°

a   PLANOGRAPHIC   b   MILITARY PERSPECTIVE

FIGURE 226 Planometric projections of the top, front, and right side. A: Planometric projection: All measurements are to the same scale as the plan; B: Military perspective: Planometric projection in which the vertical scale has been adjusted (usually by 1/2) for greater realism, as in cabinet projection.

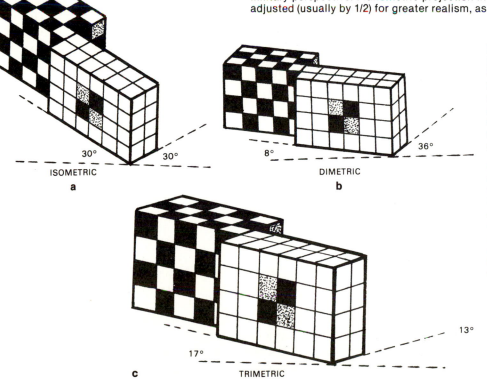

AXONOMETRIC

30°   30°   8°   36°

ISOMETRIC   DIMETRIC

a   b

17°   13°

c   TRIMETRIC

FIGURE 227 Axonometric projections of the top, rear, and right side. A: Isometric projection: All measurements are to the same scale as the plan; B: Dimetric projection: The depth ratio is 1/2 that of the front; C: Trimetric: This has a depth ratio of 2/3rds the front and a vertical ratio of 5/6ths.

Dimetric and trimetric projection are simply versions of isometric in which two of the dimensions of height, width, and depth have been adjusted to give a more realistic image. Typically, the adjustments are in the ratios given here. The angles are somewhat arbitrary, but it is quite common in dimetric to give both planes a 15° slant.

155

three kinds: *isometric, dimetric,* and *trimetric.* Of these, isometric projection is by far the most commonly encountered and, of all the modes of axonometric, it is the most useful and easily learned. Its principal advantage is that all measurements in the projection are to the same scale, yet the shape of the object represented is not as distorted-looking as it is in cavalier or planographic projection. The word *isometric* means "of equal measure" and describes Fig. 227A perfectly. *Dimetric* (double measure) and *trimetric* (triple measure) describe Figs. 227B and 227C with equal precision. Dimetric projections look slightly more plausible than isometrics simply because the viewer seems to be staring down on top of objects; the front or side faces of things can be emphasized. However, dimetric rendering is subject to the same objection as cabinet projection and military perspective. Working in two different scales is tedious. Trimetric, of course, requires three translations of scale, one for each dimensional plane, and most of us consider it far too much trouble for a very small gain, namely, the fact that renderings done this way look a bit less schematic than the other axonometric projections and afford the option of stressing the top or either side of an object. They are popular, however, with one group of professionals, who customarily refer to them as "axonometric drawings." Frequently architects represent large structures and elements of structures with trimetric projections like the one in Fig. 228. It is easier, in fact, for architectural *delineators* to do trimetric projections than to undertake the usual alternative, central projections.

Central projection is the technical name for scientific perspective. The center of this projective technique is the center of one eye—or any point of focus. In Dürer's woodcut (Fig. 221) the hook in the wall from which the string radiates out to the different points represents the eye. If we wished to do so, we could explain how the eye is where the hook is or where the camera lens would be were we to photograph the lute, and also how it corresponds to the point where the edges of the table would converge if drawn far enough beyond the lute. One does not have to understand the optical facts in order to appreciate the character of perspective drawing, however. For the purposes of this book a brief introduction is sufficient.

Just bear in mind that perspective drawing is not *the* correct way to represent things but is merely a more complicated form of engineering drawing associated with projective geometry. Perspective theory depends upon the idea of a functional infinity. Ordinary perspective drawing depends upon two fundamental notions: (1) the horizon line or *eye level* and (2)

FIGURE 228 Trimetric architectural projection. Usually referred to by architects simply as "axonometric projection." The angles are 60° and 30°. *Arresicondo,* 1982, courtesy of Arressico.

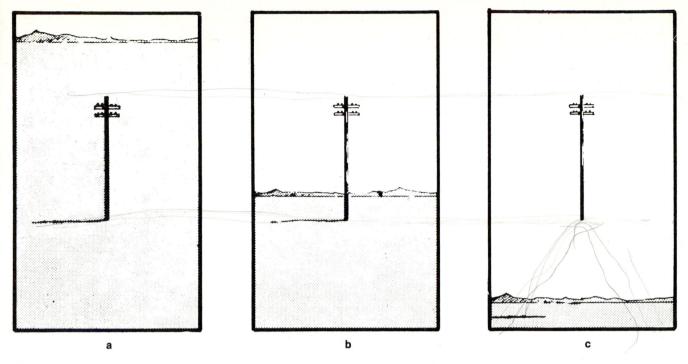

FIGURE 229 Horizon line and viewer's eye level. The horizon line in the pictures is identical with the eye level of the onlooker. The only difference among the three pictures is the position of the horizon relative to the pole.

vanishing points toward which the edges of things converge.

Fig. 229 contains three pictures of a telephone pole. In Fig. 229A we are above the pole, as if in a helicopter. In Fig. 229B we are standing looking at the pole. In Fig. 229C the pole is higher than we are—it's floating. The drawing of the pole is the same in all three pictures; the only change is the position of the line of the horizon relative to the pole. When you are above the pole, the line is above it. When you are on the ground, the horizon line passes through the pole. When you are below the pole, the line is below the pole. All of us have seen telephone poles from these angles in essentially these ways. (You would see one like 229C if it had a steep hill running up from the bottom to support it.) The horizon line is *always* and invariably at the same level as your eyes. *Always!* Even when you are flying in an airplane, the horizon line is level with your eyes. If you are looking out over the wing, you will see that the edge of the Earth or a cloud is passing through your window, not lying beneath the fuselage.

You can check the truth of this statement by imitating the people in Fig. 230. Look through a sash-hung window, a casement window, or any window with open Venetian blinds. Stand looking at the horizon so that some other horizontal (sash, crossbar,

blind slat) matches the line of the Earth against the sky. Squat. The horizon will descend. Stand on a chair. The horizon will ascend. Climb on a ladder and it will rise still higher. Lie on the floor and it will follow you down. In Fig. 230 you are given the eye level of one of the individuals in each of the cartoons. The artist lying on the floor sees the room from a different perspective than the robot who has mistaken Leger's *Three Women* for a pinup. The gallery worker on the ladder looks down on everything else. Notice that the horizon line passes through everything no matter how near or far. "Horizon line" used as we are using it refers to the edge of the globe. Trees and houses and mountains get in the way, so keep it in mind that the eye level is what we are really concerned with.

Imagine that you are standing in the desert in the middle of a railroad track (Fig. 231). Of course, the rails converge to a single point. What you might not realize is that every edge or alignment parallel to the rails is directed to the same point (Fig. 231B). This is a perspective drawing. The horizon line establishes the viewer's eye level relative to the scene. The point at which all of those diagonals converge is called a *vanishing point,* and it is what establishes the viewer's position. In Fig. 231C the viewer has moved just left of the tracks. In Fig. 231D the viewer is up on a water

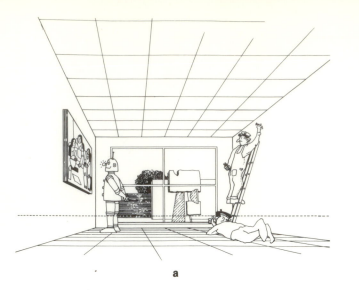

a

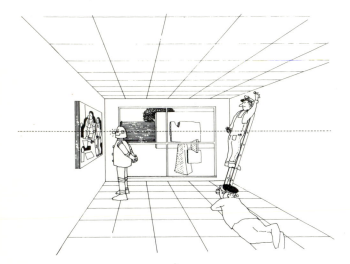

b

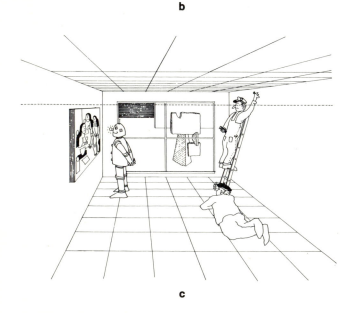

c

FIGURE 230 In each of these views of this art gallery and the landscape beyond it you, as the viewer, have the same eye level as one of the characters in the illustration. A is the artist's very low view. B is the eye level of the robot. C is the eye level of the man on the ladder.

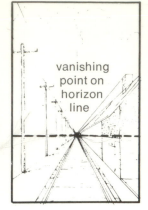

FIGURE 231A A railroad track and shed with some telephone poles along the track.

FIGURE 231B The basic lay-out in *A* with lines drawn to the vanishing point.

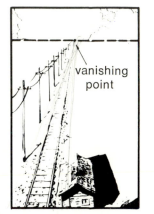

FIGURE 231C The viewer stands at left of the track. The horizon line is in the same place, but the vanishing point has shifted left.

FIGURE 231D The viewer has climbed to a higher eleva-tion to the right of the track and the horizon line has risen and the vansihing point moved over slightly to the right.

tower. This is the kind of imagery that involves only one vanishing point.

What if the viewer were supposed to be standing way around to the side of the tracks, as in Fig. 232? Where would the ties go? How do you know where to put the crossbars on the poles? In such a case the art-ist sets down a second vanishing point and directs whatever is at right angles to the rails to that point, as in Fig. 233. (As will be shown in a bit, it is possible to designate the precise spot for the vanishing points on the horizon line. However, it is usually not necessary to be so finicky; the principal thing is to keep them quite far apart. In two-point perspective the two op-posed points will never be inside the same picture space.) Actually, in Fig. 233 there is in fact a third vanishing point for the nearby crate and two more for the box over on the right. That makes five vanishing

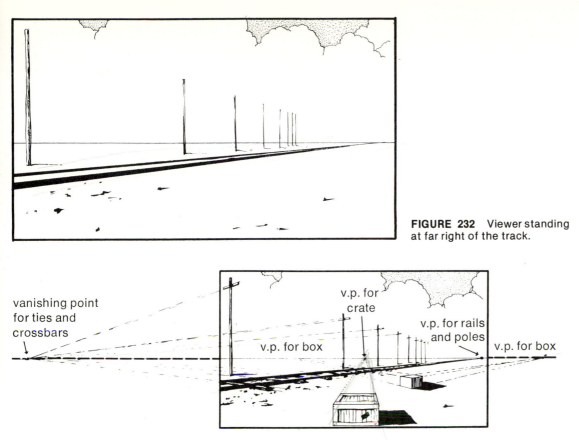

FIGURE 232 Viewer standing at far right of the track.

FIGURE 233 Two-vanishing point perspective.

vanishing point for ties and crossbars

v.p. for box

v.p. for crate

v.p. for rails and poles

v.p. for box

points altogether, but the system is called two-point perspective because two points are sufficient for rendering any one object within the picture. There is also a three-point perspective (Fig. 234), but it is uncommon in the fine arts and rarely seen even in technical illustration. In drawing certain things, such as houses with pitch roofs, there are elevated or "overhead" vanishing points (Fig. 235), but this system is included in two-point perspective.

Regardless of the number of points, any circle drawn in correct perspective will appear as an ellipse (Fig. 236) unless it is parallel to the picture plane or exactly on eye level. There are many rules and procedures for doing scientifically correct perspective renderings. The bibliography contains some superb references if you wish to learn more. In order to give some idea here of how thoroughly mathematical central projection is, we will show you how a professional

FIGURE 234 Three-vanishing-point perspective.

FIGURE 235 Sloping roof in perspective.

FIGURE 236 Circles in perspective.

v.p.

to v.p.

to v.p.

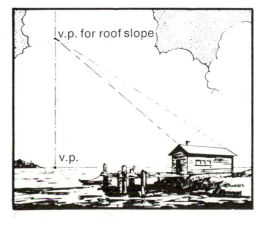

v.p. for roof slope

v.p.

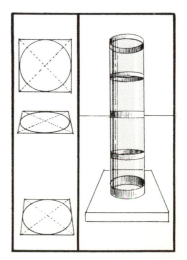

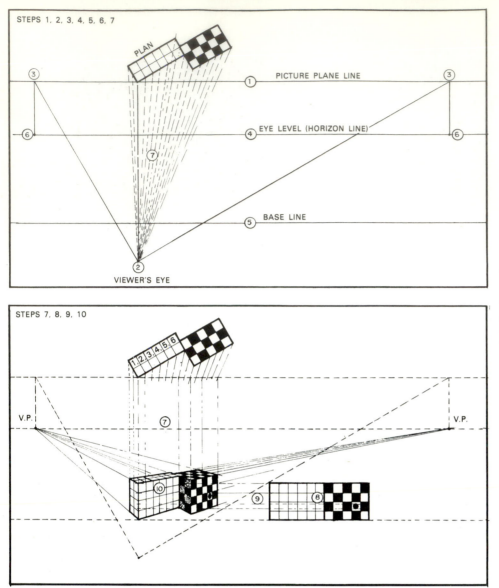

STEPS 1, 2, 3, 4, 5, 6, 7

PLAN

③ ③
① PICTURE PLANE LINE
④ EYE LEVEL (HORIZON LINE)
⑥ ⑥
⑦

⑤ BASE LINE

②
VIEWER'S EYE

STEPS 7, 8, 9, 10

1 2 3 4 5 6

V.P. V.P.
⑦
⑩ ⑨ ⑧

FIGURE 237 A perspective drawing of the stack of cubes as it would be derived mechanically from plan and elevation.

The procedure described in these diagrams will produce an absolutely accurate rendering of the stack from any angle and to whatever scale one wishes. The plan could be put down at any angle to the picture plane line and, since all the distances are in exact scale, the distance and the height of the eye level can be calculated precisely. If each of the cube faces is supposed to be a foot square, then this stack of cubes is 4 feet tall and 11 feet long. We are looking at it from 20 feet away and our eye level is just 6 feet above the stack. This is precisely the same outline a photograph taken from that viewpoint would present. It is also what a stack of 10-foot-square cubes would look like to you if you were lying on a 100-foot-high platform 200 feet from the nearest corner.

delineator would go about making a perspective drawing of the stack of cubes in Fig. 237.

When an architect shows you a perspective drawing of a building to be put up, he or she is not showing you an artist's sketch. Rather, you are looking at a drawing produced mechanically, by application of techniques of central projection. The system is so mathematical, in fact, that it can be used to program computers to do perspective drawings from any viewpoint desired. Here is how we might do the stack of cubes: (1) The plan is put down at an angle to a line representing the picture plane, with one corner or face just touching it. It could be at any angle at all, depending on what parts of the stack we wish to show. (2) A point is fixed directly below the place where the plan touches the picture plane line. This point represents the viewer's eye and will be nearer or farther from the plan depending upon how far from or near to the object the viewer is supposed to be. (3) From this point, lines are run out parallel to

the edges of the plan. (That is somewhat fortuitous, however. If the plan were trapezoidal in contour, we would still put the two lines at a 90° angle to each other with only one parallel to a plan edge.) (4) A second horizontal line is placed at some distance below the picture plane line. This is going to be the eye level and is necessarily related to the next step. (5) A third horizontal line, known as the "base line," is drawn beneath the eye level. If the drawing is to be in a specified scale and from a particular viewpoint, the distance between the lines can be calculated quite easily. (6) Directly beneath the spots where the diagonal lines intersect the picture plane line, we place two dots on the eye level. These will eventually be the vanishing points for our drawing. (7) Lines are drawn from important points on the plan to the viewpoint and marked off where they intersect the picture plane line. Then verticals are dropped from the points of intersection through the eye level. (8) An elevation is placed on the base line. (9) Measure-

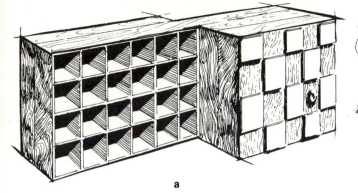

a

b

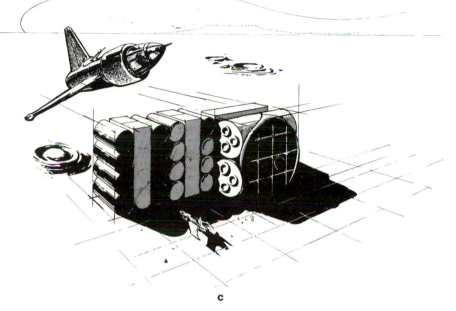

c

FIGURE 238 Variations on the theme.

Because every cube in the stack of cubes is a module, it should be easy to see how the procedure described in Fig. 237 could be used to create a drawing of a piece of office furniture (A), a building—here seen from a low vantage point—(B), or a make-believe spaceport (C).

ments on the elevation are carried across to the line connecting the viewpoint to the plan. Using these coordinates, (10) the delineator can make a perspective drawing of any object from blueprints alone. Also the angle of the plan, the height of the eye level, and the distance of the viewer from the subject can be varied to produce whatever version of the object is desired. The versions can, in fact, be quite far removed from the original (Fig. 238).

Of all forms of projection, perspective is the most nearly like human vision. Wherever it has been introduced, it has tended to drive out other ways of seeing the world for artists who wish to portray objects in a realistic way. Once you have been conditioned to the camera's eye, no other kind of drawing looks altogether right. All the same, perspective space is a very special and unnatural kind of imagery; it depends upon substituting calculation and magnitudes for the intuitive comprehension of space. Interestingly, the other kinds of projections, besides

having their geometrically exact versions used in engineering drawing and technical illustration, have their intuitive corollaries in the drawings of children and in the history of the fine arts. Observe the terrace in Leng Mei's eighteenth-century painting (Fig. 239). It very much resembles a dimetric projection. We can find many things of the same sort in the West as well. For example, Cimabue's throne looks quite a bit like two cabinet projections, one for the left half of the picture and one for the right, joined together behind the Virgin (see Fig. 219).

The first appearance of perspective in art, so far as we know, is in the frescoes of Masaccio (1401–1428), whose *Tribute Money* (Fig. 240) contains all the features of scientific perspective. The system was not invented by Masaccio, though: it seems to have been the invention of the architect Filippo Brunelleschi (1377–1447).[1]

Renaissance painters, from the time of Masaccio on, made constant use of scientific perspective. It affected them powerfully because it seemed to be the ultimate example of an abstract, theoretical system that revealed visual truths. Out of mathematical order, illusion is produced. The orderliness of perspective's completely unified space appealed to them almost more than its realistic effects. A superb example of the way in which Renaissance masters made use of the system Brunelleschi had given them is the famous *The Last Supper* (Fig. 241) by Leonardo da Vinci (1452–1519). This work is remarkably symmetrical. The twelve apostles are assembled in four groups of three, with two groups to the left of Christ and two groups to his right. Christ is posed and drawn so as to form a pyramid. He is the most stable shape in the entire work, and the most isolated and self-sufficient figure. The moment is just after He has spoken the words, "One of you shall betray me." The party is alive with speculation, and Judas, fourth from the left, draws back in fear and hatred. Of the disciples, only he has a face in shadow. He is separated from Jesus by John, whose side parallels his own. Of course, Christ is the focus of the tale and He is, quite literally, the focus of the picture. His forehead is the spot where the vanishing point resides. There the lines of ceiling converge and the tops of the tapestries on either wall are in line with His forehead (Fig. 242). Too, there are linear developments through the disciple groups leading us

FIGURE 239 LENG MEI. *Lady Walking on a Garden Terrace.* 18th century. 3′6″ × 1′10″. Museum of Fine Arts, Boston. The treatment of the terrace is somewhat like dimetric projection even though magnitudes become smaller as things get farther away from the picture plane.

FIGURE 240 MASACCIO. *The Tribute Money.* c. 1427. Fresco. Brancacci Chapel, Sta. Maria del Carmine, Florence. (Alinari-Scala)

FIGURE 241 LEONARDO DA VINCI. *The Last Supper.* 1495–98. Mural. Sta. Maria delle Grazie, Milan. Photo: Italian Government Travel Office

FIGURE 242 Leonardo's *Last Supper* showing the perspective structure and some formal properties of the composition.

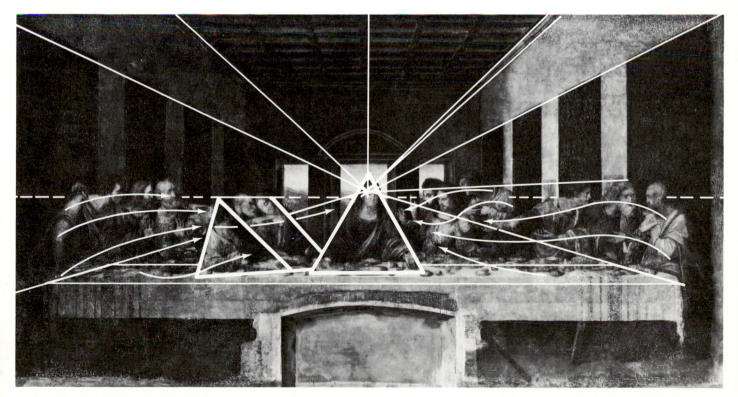

to the central figure. It is a very austere, harmonious, and thoughtful work.

Leonardo's *The Last Supper* is the most famous *Last Supper* of them all, despite the fact that he experimented with a medium that failed to adhere to the wall and that what we see in our reproduction is a flaked ghost of the original. Nearly a century later, the Venetian painter Tintoretto (1518–1594) undertook the same theme on a similar scale. His conception of the event, however, is quite different (Fig. 243). The main point of convergence is no longer Christ's face. It is over to the far right, in a remote edge of the picture. The whole thrust of the room is opposed to the viewer's glance, which is yanked across the picture toward Jesus by, for the most part, the Saviour's bright halo. This halo not only is the largest and brightest of those in the smoky, dark inn, but is also casting a light so powerful that sharp shadows fall from it across the figures in the forefront of the picture, creating very distinct pathways from the foreground to Christ. The space of the picture is powerful and convincing, but it results from employment of a distorted one-point perspective. The viewer is conceived of as standing over on the right, more or less in line with the nearest male figure. If the table were drawn in two-point perspective, as we would expect, the viewer would seem to be over to the left. Since the ends of the table are parallel to the eye level while its sides converge on the same vanishing point as other major elements of the room, we know that the viewer is remote from the center, in line with the vanishing point. That's because in one-point perspective the viewer is always directly ahead of the point of convergence. Tintoretto's aims were not scientific but emotional, and it is to his advantage to break the rules. His "errors" are partly concealed by the glamorous chiaroscuro he employs.

Usually in the fine arts the absence of scientifically incorrect perspective is neither deliberate nor the result of ineptitude. It is the consequence of an antipathy for mechanical exactitude. Edgar Degas was actively interested in perspective drawing, but his mature work (Fig. 244) is full of inconsistencies. He takes in too much of the floor; he paints things from different points of view; his receding parallels don't have a common vanishing point. *Foyer of the Dance* contains all of these imperfections. Yet the painting doesn't look unreal; on the contrary, it gives an impression of considerable accuracy. Degas's perspective is not so much inaccurate as it is empirical.

FIGURE 243 TINTORETTO. *The Last Supper.* 1592–94. Oil on canvas, 12′ × 18′8″. S. Giorgio Maggiore, Venice (Alinari-Scala)

FIGURE 244 EDGAR DEGAS. *Foyer of the Dance.* 1872. Oil on canvas, 12½ × 18″. The Louvre, Paris.

He knew the rules and they assisted him in his intuition. However, it was an impression of the room he wished to project onto the canvas and not a geometric diagram. His perspective is *sensed* rather than constructed; it is "felt" in much the way the light and shade are felt. Most artists' drawings are done freehand in the way Degas drew this picture.

Pieter Bruegel the Elder (1525/30–1569), in one of the most famous vistas in world art, his *Return of the Hunters* (Fig. 245), did not depend upon scientific perspective for the effectiveness of his spatial illusion because the landscape contains few elements that are geometrically regular. Of course, the eye level and a very firm sense of viewer position are confirmed by relationships that take perspective for granted. But the powerful impression of depth relies upon *atmospheric perspective*. This form of perspective is only incidentally related to the blurring and refractive effects of air. What it really has to do with is value contrast. *What is high in contrast appears closer than what is low in contrast.* That is, the difference between lights and darks is greater up close than it is far off. Thus, in the Bruegel, the trunks and branches of the nearby trees are far darker than distant ones, and those in the middle distance are darker than the ones very far away. At the same time, the foreground snow is whiter than that in the valley, and the snow of the valley is not so dark as that on the mountainside. Everything gets grayer toward the horizon. What is important to the illusion, however, is the decrease of contrast *between* lights and darks. Things might grow darker, leaving an impression of gloom; things might become lighter, as if things far off were lost in luminous haze. Con-

FIGURE 245 PIETER BRUEGEL THE ELDER. *The Return of the Hunters.* 1565. Oil on panel, 46 × 63³/₄″. Kunsthistorisches Museum, Vienna

trast between things, then, is the key to atmospheric illusion (see Fig. 246).

The cases here cited, those using combinations of perspective drawing and atmospheric perspective, are traditional treatments of space in painting. However, there are other kinds of spatial treatments that have had an immense effect upon the art succeeding them. One of the most important and least understood is the space invented by Paul Cézanne (1839–1906) and extended by Cubism.

It is easy to find misinformation about the space of Cézanne since it is constantly being published de-

a

b

c

d

FIGURE 246 Atmospheric perspective, absent and handled in three different ways.

Here is a landscape in which the illusion of depth depends, in Fig. 246A, exclusively upon overlapping and relative scale. When the light areas are made progressively darker as they recede into the distance (B), an increased illusion of depth is achieved. One might also suggest depth by (C) lightening the darks and leaving the light areas alone. In D the illustrator has done as Bruegel did and made the lights grow darker and the darks grow lighter.

FIGURE 247 PAUL CEZANNE. *The Basket of Apples.* 1890–94. Oil on canvas, 25¾ × 32″. The Art Institute of Chicago. Helen Birch Bartlett Memorial Collection

spite the tireless efforts of a few to correct the errors. One of the most common falsehoods, and among the easiest to disprove, is that the artist based his designs (Fig. 247) on the reduction of things to ideal geometric forms of cones, cubes, spheres, and cylinders. People who should know better believe so because of something Cézanne wrote to younger artist Emile Bernard in April 1904. However, he did not say what he is purported to have said. This is the *actual* statement: "May I repeat what I told you here; treat nature by the cylinder, the sphere, the cone, everything in proper perspective so that each side of an object or a plane is directed towards a central point."[2] You will notice that he failed to mention cubes —the form most commonly associated with his style—and that he *does* definitely say that everything should be "in proper perspective," a condition to which his own paintings are immune. One has only to look at his work to see that the shapes in them are not especially geometric. That is, in *The Basket of Apples* the real apples were probably at least as spherical as the painted ones. The most obvious thing about the objects in this still life is that they are not in perspective. Typically, Cézanne "blunts" convergences and flattens out ellipses. Why, then, his instruction to Bernard? A second letter, this one to his son Paul in 1906, gives us a good clue. The master mentions "the unfortunate Emile Bernard . . . an intellectual engorged by the memory of museums, but who does not look enough at nature."[3] In the earlier epistle, Cézanne was not telling Bernard the "secret" of how to paint a Cézanne; he was instructing him on how to look at nature—in terms of basic geometric forms. This was standard academic advice to students at the time and, indeed, is still standard advice to beginners today.

Cézanne's style depends upon two things. The first is his deviation from proportions like those in Figure 248A, proportions having an existence in the rules of perspective which dictate how the subject is to be drawn. A schematic of the Cézanne version of the bottle and apples on the table reveals how much more interrelated all of the pieces are (Fig. 248B). The apples are like the cloth in their generalized lumpiness, the outline of the cloth like the table itself. The bottle and the table, which are both asymmetrical, being different on their left and right sides, are unlike the projected bottle and table in Fig. 248A. Cézanne's objects have no prior existence in theory or in fact. His achievement of a final order is therefore a special attainment of his own.

The second feature of Cézanne's style is associated with the first and is usually referred to as his "perspective of color." This is a very complicated matter and depends upon properties of color relationships much more involved than anything so simple as application of the principle that "warm colors advance and cool ones recede." That principle is for the most part true, but there are plenty of exceptions to it. As a matter of fact, the nearest things in Cézanne's still life are among the coolest in hue. Suffice it to say that Cézanne's drawing and color, in concert, create a massively spacious assembly of forms. He has married classical structure and harmony with Impressionist color in such a way that his art contains both and resembles neither.

The unique spatial and compositional properties of Cézanne's art fascinated young turn-of-the-century artists whose training was traditional but whose temperaments thrilled to the adventurous promise of unconventional directions. Two of the most talented of these, Pablo Picasso and Georges Braque, invented the style that was to revolutionize painting in modern times. It came to be called Cubism, although it had no more to do with cubes than Cézanne's art had to do with cones and cylinders. It also had very little to do with what have been called "multiple viewpoints" and "multiple

a

b

FIGURE 248   A: Cezanne's subject-matter treated in a traditional manner; B: Schematic rendering of the Cezanne still life.

perspective," although it is possible to find what might be considered combinations of separate views in a single picture. Thus, Picasso's *Girl Before a Mirror* (Colorplate 14) shows us a profile of the girl's face superimposed upon a full-face view. Such combinations of forms, however, are not common in Cubism except among lesser artists before the late twenties, and they tend to appear in the work of Picasso and Braque with some frequency in 1929 and later.[4]

We have already analyzed the spatial complications of Picasso's *Portrait of Ambroise Vollard* (Fig. 249) in Chapter 4. Now, consider how these intricacies relate to the monocular space clue of primitive overlapping we examined at the beginning of this chapter. The schematic drawing in Fig. 250A sets the conditions for 250B. However, it is obviously not as easy to be clear in one's mind that *Q* really is behind the

FIGURE 249 PABLO PICASSO. *Portrait of Ambroise Vollard.* 1909–10. Oil on canvas, 36 × 25½″. Pushkin Museum, Moscow.

others in Fig. 250B. Where is *P*? Possibly it lies both ahead of and behind the other forms and lines. This effect is clearly not due to some transparency of the forms. Neither is the ambiguity of elements in the *Portrait of Ambroise Vollard.* In both cases the uncertainty emerges from the arrangement of masses and edges of masses. The bizarre space of Cubism is of great interest because, as we noted earlier, the whole character of a painting like this one by Picasso changes as the viewer changes assumptions about where in space the elements rest.

In 1912 Cubist collages reinstated the unbroken area and, in paintings done subsequently, the treatment of zones as solid units was continued. Still, the spatial relationships did not change in any great way. Picasso's *La Suze* (Fig. 251) is just like our diagram in Fig. 250B so far as its internal relationships are concerned.

Cubism's uncanny play of forms in space opened up a vast new range of possibilities for creative artists. During the course of this century we have had to learn to deal with images that would have been meaningless in the nineteenth century even to people as sophisticated as Cézanne. Aerial photography (Fig. 252) and microscopy (Fig. 253) are replete with fantastic images. Furthermore, contemporary artists have taken it as part of their job to challenge viewers with art that falls beyond the "safe range" of easily understood visual material provided by such preordained, logical systems as orthographic projection. In Hans Hofmann's *The Golden Wall* (Colorplate 3) we have an example of an artist playing with space.

FIGURE 250 A: Overlapping planes; B: Cubistic treatment of space

Traditional overlapping space contrasted with the space of Cubism. There is nothing to prevent us from assuming that the overlapping in Fig. 250B is exactly the same as that suggested by Fig. 250A. On the other hand, it is simple enough to suppose that Q is the uppermost form in Fig. 250B, not so easy to make this assumption about Fig. 250A.

a b

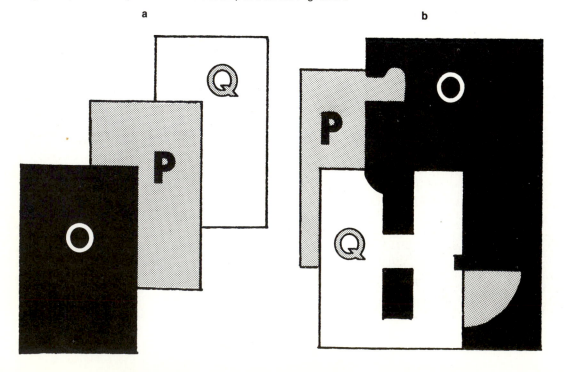

FIGURE 251 PABLO PICASSO. *La Suze.* 1912–13. Pasted paper with charcoal, 25¾ × 19¾″. Washington University Gallery of Art, St. Louis

The Golden Wall is made up of roughly geometric shapes in complementary hues, hanging like panels from a ceiling, some of them overlapping other zones, some seeming merely adjacent to others. In a number of places the cooler hues are made to come forward by reason of the context in which they are encountered. Thus, the blue rectangle at the bottom advances beyond the hot reds because it is seen as part

of the same family of colors as the azure above and the blue-violet strip seen against yellow on the left. The comparative brightness of the violet, blues, and lime greens pushes them toward us. The magenta hangs in an uncertain space between the yellow rectangle and the hot red-orange blocks in the upper middle of the painting. The relative scale of the green areas functions in this context to suggest that the smaller unit is quite far back. Surely, it appears to be overlapped by the blue, but is it behind the magenta strip or ahead of it? The blobs of color spotted about here and there are important to the game Hofmann plays because they act like supporting chords in a musical composition. For example, the small touches of yellow and brown along the top edges of the ultramarine blue square help position the square in space. The narrow aisles between the rectangles clustered in the upper right "squeeze" space into narrow channels and produce a lot of tension between edges and forms. They not only give the comparatively small area the power to balance massively predominate reds, they also focus attention on the margins of the work. The larger green unit is not quite segmented by the frame, but the yellow one and the blue on top left are, implying the indefinite continuation of Hofmann's picture space in the way so typical of modern art from Monet on.

The Golden Wall is spatially tantalizing. Hofmann is performing his prestidigitator's game of "push-pull," as he called it, by setting the surface into contradiction with the space, playing off against one another contradictory monocular space clues such as scale, hue temperature, and overlapping. The subtle elusiveness of the thing is such that even after continued viewing the color swatches seem to sneak in and out of rein, challenging our ability to bring these mischievous forms into any kind of final resolution. That we cannot is, of course, the most delightful aspect of the work. It demands solutions from

FIGURE 252 Airview. Southern Illinois University at Edwardsville. Photo: John Rendleman

FIGURE 253 Microspectroscopic photograph. Photo: Frank B. Kulfinsky

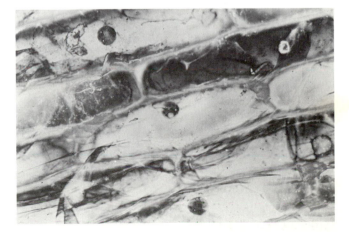

FIGURE 254 DENNIS RINGERING. *Gold Standard.* 1980. Acrylic on canvas, 44 × 54″. Private collection

us instead of supplying them ready-made the way traditional art usually does.

Dennis Ringering (b. 1942), to take a very different kind of artist, uses conventional devices of atmospheric perspective to produce a similarly disconcerting illusion in his *Gold Standard* (Fig. 254). Here, the sharp, clean bar casts a harsh shadow upon a

background that seems ripped by contrasts, floating like islands of distance in a cloudless sky. The terminal coincidence of two of the rents with the graduated stripe further confounds the relationship of one surface to another. These uncertainties, however, seem far removed from the ostensibly inflexible "gold standard" of the title.

The title of Ringering's airbrush painting relates to the Hofmann painting only by chance, but its satirical comment upon things quantitative and qualities unmeasured is characteristic of the reigning contemporary artistic sensibility which sees in the rationally objective always some faint suggestion of the arbitrary. Michael Smith's calligraphic studies express a sense for discovering in what is intuitive and poetical a certain measure of calculated irony. *Traveler "K"* (Fig. 255) seems to hurdle from the boundaries of the pictorial into the world at large. An earlier series of paintings featured carefully ruled vertical bands against nebulous blobs of color. In *Crimson Binary* (Fig. 256) gray strips overlap alizarin cumulus. In *Cobalt Binary* (Fig. 257) the strips are pale green against dark green stains. They are against the stains; they don't overlap the white areas. Somehow we expect the interruption to be the other way about if it is going to occur at all. So here we are forced to play with the positive–negative features of figures and grounds not because of ambiguity but because of an unexpected prejudice in favor of simple forms being presented to us in a certain way. Richard Duncan (b. 1944) in his print (Fig. 258) doesn't defy the bias; he overwhelms it with complications.

FIGURE 255 MICHAEL SMITH. *Traveler "K".* 1978. Oil on paper, 19 × 20″. Private collection

FIGURE 256 MICHAEL SMITH. *Crimson Binary.* 1971. Oil on paper, 19 × 22″. Arressico collection

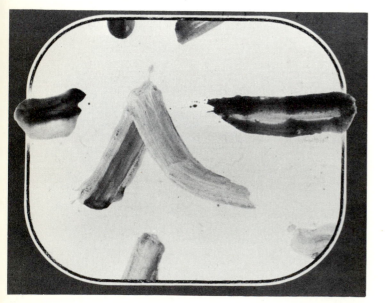

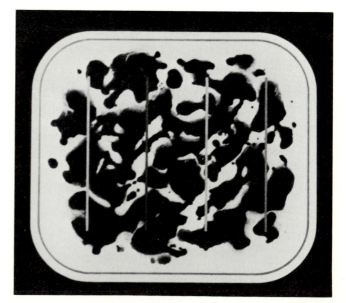

FIGURE 257 MICHAEL SMITH. *Cobalt Binary.* 1971. Oil on paper, 19 × 22″. Arressico collection

FIGURE 258 RICHARD DUNCAN. *Horizontal Object.* 1977. Etching, 17 × 34″. Private collection

An intriguing merger of conventional delineation of three-dimensional forms with an avant-garde view of art as being an imaginative challenge rather than a certain category of objects or activities occurs in the work of Conceptual artist Peter Eisenman (b. 1932), an architectural theorist. The plan of *House IV* (Fig. 259) is not intended to represent a building; instead, it is a set of permutations which pertains to orderliness and dimensionality of objects in the same way arithmetical measurements do. The trimetric projection of a rectilinear block first seen in the upper left undergoes various changes, each based upon the one behind and the one above any given unit. The drawings are ''to be read two ways simultaneously. Horizontally, *House IV* shows the series of systematic transformations the basic space undergoes. Read vertically, the drawing provides translations of these steps into formations of planes and columns.''[5]

Eisenman's ''houses'' are not really three-dimensional, but they correspond to primary rela-

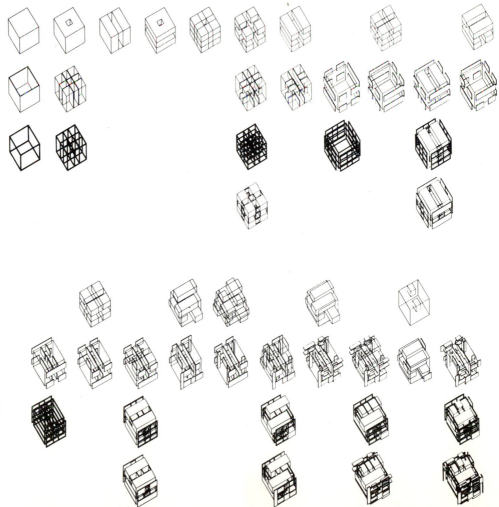

FIGURE 259 PETER EISENMAN. *House IV.* 1971. Ink and white paint on clear mylar film, 51 × 17″. (Delineator, Robert Cole, 1975)

tionships which do exist in height, width, and depth. Space is obviously actual as well as pictorial. Sculpture and architecture are the most clearly spatial of the fine arts simply because, as physical entities, works of sculpture and architecture usually exist in the third dimension. Thomas Gipe (b. 1938), a sculptor, blacksmith, and graphic artist, brings very simple volumes into impressive alliances. His sculpture is like one of Eisenman's trimetric diagrams in suggesting a metaphor for systematic transformations. *Cordoba Brown DeKalb* (Fig. 260) looks like some secretly organic mechanism that is in the process of

FIGURE 260 THOMAS GIPE. *Cordoba Brown DeKalb.* 1980. Welded steel with lacquer finish, 31″ tall. Arressico collection

FIGURE 261 MARK DE SUVERO. *XV,* 1971. Steel, 21′7″ × 26′11″ × 23′11″. Laumeier International Sculpture Park, St. Louis. Courtesy of Oil and Steel Gallery, New York.

FIGURE 262 RICHARD HUNT. *Extending Horizontal Form.* 1958. Steel, length 4′9″. Whitney Museum of American Art, New York (Gift of the Friends of the Whitney Museum of Art)

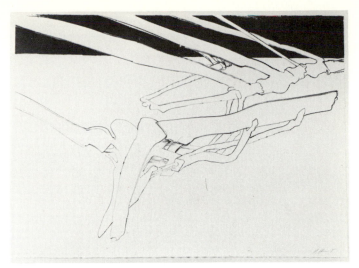

FIGURE 263 RICHARD HUNT. Untitled. 1972. Lithograph, 22 × 30″. Collection Southern Illinois University at Edwardsville.

FIGURE 264 THOMAS GIPE. *Interior with Sculptured Objects.* 1980. Pencil drawing, 14 1/2 × 19 1/2 ″. Private collection

evolving into a more advanced gadget. From a simple box, tubes sprout and sharp edges soften into regulated curves. Even the softly glistening skin of metallic lacquers, so evocative of the modern world of sleek, shiny automobiles and plastic lounges and synthetic dreams, elicits thoughts of change, mobility, and unfathomable potential.

Gipe's sculpture is very compact. It contains space within its forms, confines a distinct volume, and concentrates it so that what is perfectly apparent seems somehow secret. Sculptors Mark di Suvero (see Fig. 261) and Richard Hunt (b. 1935) are identified with sculpture that inhabits spatial voids. Hunt's structure of rods, lumps, and panels (Fig.

262) is a three-dimensional drawing in which lines have been made from steel. Like Suvero's gigantic constructions, Hunt's openwork designs reveal perpetually new and unexpected relationships as one examines them from different viewpoints. His elegant, rather menacing shapes remind one of macroscopic insects and volatile vegetation. Hunt's drawings, too, resemble fragments from the universe of the microscopic slide (Fig. 263), segments of a greater whole than we know seen in momentary isolation against a void. Gipe's drawings, by contrast, show us his rather odd shapes in a conventional space (Fig. 264). His interior is a spatial container, like the sculptures resident within it.

EXERCISE 8–1: Denying Optical Truth

Materials: White drawing paper
Tools: Pencils and straight edge

Eighteenth-century satirist William Hogarth (1697–1764) once did a humorous engraving (Fig. 265) attempting to defy the conventions of perspective. He did not make the world look flat or the space appear completely distorted because he retained (perhaps by inadvertence) the general notion of a viewpoint. He did, however, manage to produce a very strange set of goings-on within the space before him. Study the reproduction and enumerate as many of the "mistakes" as you can. Then undertake to emulate Hogarth's experiment by doing a simple outline drawing of the room you are in, insinuating into this drawing the same kinds of errors Hogarth did. He has, for instance, reversed the expected

FIGURE 265 WILLIAM HOGARTH. *Frontispiece to "Kirby's Perspective"* (Joshua Kirby's edition of Dr. Brook Taylor's *Method of Perspective*). 1753. Engraving, 8¼ × 6¾". Reproduced by permission of the Trustees of The British Museum, London

diminution of more distant things in his rendering of the herd of sheep. He brought vastly separated levels of space onto the same plane, enabling the lady in the window to light a pipe for a man standing on a distant hilltop. Nearer objects are overlapped by more distant ones. Things converge in the wrong directions. We can see rooftops as if we were looking down upon them even as we see lower parts of the same buildings as if from below. The bridge doesn't sit level on the water. And so on.

Doing things wrong can be amusing, of course. It appeals to our sense of the perverse. It can also be instructive, for it makes one conscious of what may otherwise be taken for granted.

EXERCISE 8–2: Frontalizing Deep Space

Materials: White drawing paper
Designer colors (or colored papers and rubber cement)
Picture magazine
Tools: Appropriate to medium used

Hogarth's space is not flat. It isn't even what we might call *frontal*; it's the space of illusion treated improperly. Picasso's *Girl Before a Mirror* (Colorplate 14), though, is what we think of as being a *frontal* treatment of space. In the terms used earlier in this chapter, it focuses our attention on the screen rather than on the world beyond the screen. Select from a picture magazine a photographed scene containing buildings. Paste it down onto a sheet of white drawing paper and, using a straight-edge and ball-point pen, draw lines along the major edges as has been done in Fig. 266. Photographs do not always come out perfectly correct—especially after they have been reproduced in magazines—because there are many opportunities for tiny deformations to occur during enlarging, in the course of making printing plates for presses, and in printing itself. Too, the photographer may have used special attachments to "correct" the original lens-eye view of the site. For the most part, however, photographs of rectilinear objects will be in correct scientific perspective.

Now take the photograph and redesign it so that it is no longer three-dimensional in effect. Compose an alternative representation deleting the depth-creating forms and emphasizing the pattern-creating ones. You may do this as a painting or as a collage enhanced with some drawing. Fig. 267 is a painted example of such a modification.

FIGURE 266 Photograph with lines superimposed to show perspective. Dotted line indicates eye level of slightly tilted camera.

FIGURE 267 "Frontalized" version of the photograph in Fig. 266. This design is not intended to look completely flat, but is intended to emphasize planes parallel to the picture surface.

FIGURE 268 Metamorphosis of a projection

EXERCISE 8-3: The Metamorphosis of a Projection

Materials: White drawing paper
India ink
Tools: T-square and triangles
Pencils
Ruling pen or technical fountain pen
Ruler

This exercise will give some experience in doing delineation, using an actual object instead of a plan.

Secure a rectilinear object of modest size—for example, tissue box, book, drawer, or carton—and measure it. Then do a drawing in oblique, planometric, or axonometric projection of the object at one-half or one-quarter scale. If the object is 12″ × 8″ × 4″, make the drawing 6″ × 4″ × 2″ or 3″ × 2″ × 1″. (If you choose to do a cabinet projection or some other projection in which the scale of different faces is not the same, you must adjust accordingly.)

Next, use the drawing as the basis of a series of metamorphoses of the shape into some other shape. This exercise has something in common with the Peter Eisenman study (Fig. 259). Fig. 268, done by a student, transforms a gift box into a Thomas Gipe sculpture.

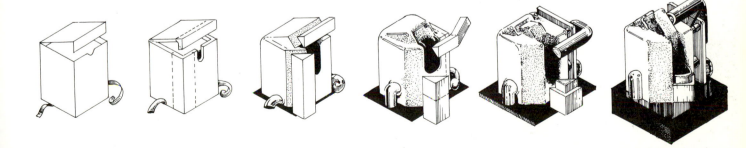

EXERCISE 8–4: Overlapping and Perspective

Materials: White drawing paper
Picture magazines
Rubber cement

Tools: Pencils
Technical fountain pen or ruling pen
Single-edge razor blade or frisket knife

Cut some shapes from magazine reproductions, getting a variety of sizes, hues, intensities, and values. On a sheet of white paper draw a horizon line. Then arrange your scraps so that they overlap to convey a sense of great depth. Consider the influences of atmospheric perspective as well as scale relationships. Generally, things high in contrast will seem nearer than those low in contrast. You will also find that generally (but not always) brighter colors tend to advance and light values are thrust forward by dark backgrounds.

EXERCISE 8–5: Small-scale Sculpture

Materials: Model-builders' balsa wood stock
Scrap mat board or illustration board, etc.
Miscellaneous materials such as thread, toothpicks, pins

Tools: Frisket knife
Pencil
Scissors
Whatever else is required by project

The pictures in Fig. 269 are of small sculptures made by students over a number of years. The styles

FIGURE 269 Student small scale sculptures

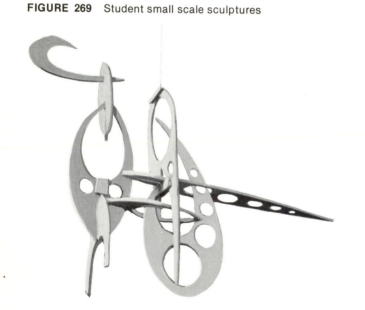

and concepts vary enormously, as do the materials, but the sculptures are all under 8 inches tall and are made mostly from materials readily available in hobby shops and art stores. Most of them could be fabricated in metal or wood on a far larger scale.

In building a work of this sort, try to consider every point of view and the way each relates to the next. Color can, of course, be an important feature of abstract designs; the hanging piece with all of the pointy elements was finished off in bright orange.

notes for chapter 8

[1] The first record of perspective theory appears in a little book published after Masaccio's death. It is Leonbattista Alberti's *Della pittura,* the most influential of all early Renaissance treatises on art. It first appeared in Latin in 1435 under the title *De pictura* and in the following year was rewritten in Italian by the author, who dedicated it to Brunelleschi.

[2] Paul Cézanne, *Letters,* ed. John Rewald, trans. Marguerite Kay (London: Cassiver, 1941), p. 234.

[3] Cézanne, p. 268.

[4] Cézanne and the Cubists have continually suffered the results of an ''explanation'' of Cubism put forth by two lesser Cubists, Jean Metzinger and Albert Gleizes, in 1912. They maintained that Cubist designs resulted from deliberate fragmentation of the world into different viewpoints which the artist then reassembled as separate fragments into a picture. They also suggested that Cubism thereby made artistic statements which were somehow parallel to the space-time world of Einstein's Relativity. Over and over again, knowledgeable critics have demonstrated that the belief in Cubism's multiple viewpoints is without merit and that, in any case, Cubism has nothing at all to do with the Theory of Relativity. But the cliché seems to be ineradicable. See Fig. 87, caption.

[5] Susan Korb, *Numerals: 1924–1977,* an exhibition catalogue, ed. Rainer F. Crone. Yale University Art Gallery, 1978. Unpaged.

part **4**

applications

chapter 9

graphic applications

Design is a general term, comprising all aspects of organization in the visual arts. In this book we have tended to emphasize the general adhibition of design fundamentals rather than their special utility, but certain activities are more commonly associated with design than others. When someone is referred to as a "designer," most people rarely imagine that he or she is a sculptor, painter, or architect. More often they assume that the person is involved with printed imagery, interior decor, or haute couture. Of occupations, graphic designing is surely the most obvious practical application of the knowledge dealt with in this book. Since even laypersons live surrounded by printed images—indeed, are induced by them to buy certain kinds of furniture and garments—it seems reasonable to devote some space to the design of such imagery.

Most people are quite astonished when they have an opportunity to compare the actual art work for an advertisement or magazine cover with the printed version. In the first place, the original drawings and lettering are usually up to twice the size of what appears in print. Where the photograph of the product appears on the page, there's a solid black square or a red outline on the art work. Usually, the colors show up only in the printed version. Everything is in black and white, except, perhaps, for pale blue pencil lines. The designer may have painted out areas with white paint or made corrections by pasting one thing over another. In print, none of this will be evident. The image will be smaller, sharper, cleaner, and clearer. Some idea of the difference between the materials a designer submits and what the printer does with them can be had by comparing and contrasting Fig. 270 with the result in Fig. 271. Such

contrasts are characteristic of nearly every graphic production, and a special jargon has grown up to describe the various elements and phases of the process that takes a project from the artist's studio to the printed page.

The prepared art work, such as that in Fig. 270, is known throughout the industry as a *mechanical*. A mechanical may consist of something as simple as a pen and ink drawing or it may have many overlays carrying registration marks and elaborate instructions, photographs to be incorporated in the design, and manuscript to be composed in the form of type. Any material intended for reproduction is called *copy* (whether it is manuscript to be set in type or a picture). The copy in the form of pictorial material will be photographed to make a film *negative*, which, in turn, is used to create a printing *plate*. From the plate a press will print an image onto paper *stock*. That image is known to printers as an *impression*. Any copy that requires no real modification apart from reduction or enlargement of scale to match the desired impression is known as *camera ready*. That is, nothing is required for a plate to be prepared apart from the production of negatives.

A designer's concern is with preparation of mechanicals—using his or her skills to articulate letterforms created by typographers, art work done by illustrators and photographers, and techniques available to technicians known as *platemakers*. The requirements for doing mechanicals are technical rather than artistic, and they are too involved to give more than cursory mention in an introductory text of this kind. However, you should know something of the process that turns mechanicals into impressions.

FIGURE 270 Mechanical for wraparound magazine cover

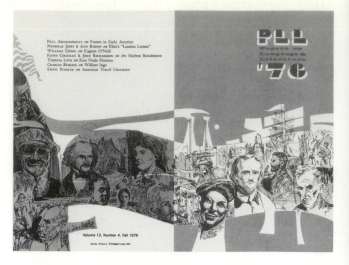

FIGURE 271 JOHN ADKINS RICHARDSON. Magazine cover. *Papers on Language and Literature* (Fall 1976). Southern Illinois University at Edwardsville

The actual art works from which graphic designs result usually do not closely resemble the finished printed impression. The material in Fig. 270 is the mechanical for the finished magazine cover in Fig. 271. The term *mechanical* derives from the kinds of drawing instruments used in preparing the paste-ups and overlays. (The word *keyline* is synonymous with *mechanical;* most paste-up art contains "key lines," outlining areas where a tint or halftone is to be positioned.)

FIGURE 272 A design incorporating grays in the form of black dots on white paper

photomechanical reproduction processes: letterpress _____

Except for the colorplates, all of the pictures in this book were reproduced by a process that prints solid black onto white paper. Whenever ink touched paper, there is a spot of black, and where ink did not, the paper remains white. There are no in-betweens, no grays of any kind. Fig. 272 does seem to contain zones that are gray, but if you look closely, you will observe that such grays are made up of tiny black dots distributed over areas of white and that, while the op-tical effect is of gray, the physical fact is black on white. This is important to understand because the two most common printing processes depend upon it.

Let's suppose that we wish to reproduce a little drawing like that in Fig. 273A. It is a very simple mechanical posing no problems. The printer can use it as copy; it is what printers call *camera ready.* The printer's camera is probably unlike any you have seen before. It is a *process camera,* and even small ones are quite large. Fig. 274 shows it in its essential detail. The copy (Fig. 273A) will be "shot" through the camera and the negative used to process a plate. As you can see from the caption to Fig. 273, what results in this case is a zinc photoengraving called a

FIGURE 273 Various stages in the creation of a line cut

The drawing in Fig. 273A is a simple example of camera-ready line copy. It will be centered on the copyboard at the far end of a process camera pressed flat against it by banks of arc lamps set at 45° to eliminate shade or glare. When the film in the camera has been exposed and developed, the result is a negative (Fig. 273B) of the original drawing, in this case reduced to 50 percent of the original size. What was black in the drawing is transparent (not white) in the negative. The next step uses the negative to make a surface from which ink can be transferred to paper. In letterpress the negative is *flopped*—turned over to make a mirror image— and (Fig. 273C) mounted on a piece of plate glass called a *flat,* which is placed face down onto a sheet of zinc that has been coated with a photosensitive solution. It is then exposed to very strong arc light. The black of the negative (white in the drawing) blocks out the light; the clear area of the negative (black in the drawing) permits light to pass through, where it hardens the coating on the zinc so that it is insoluble in water (Fig. 273D). After the unexposed areas have been washed away, the exposed areas are made impervious to acid and the plate of zinc is immersed in a solution of nitric acid. The areas that were white in the drawing are eaten away, leaving the areas that were black in the drawing standing up in relief. Mounted on a wooden block (Fig. 273E) to be brought up to the same height as printer's type, the plate is printed exactly like type. What it prints is a small duplicate (Fig. 273F) of the original drawing.

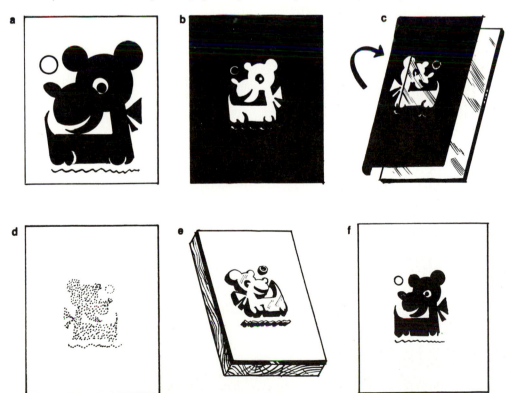

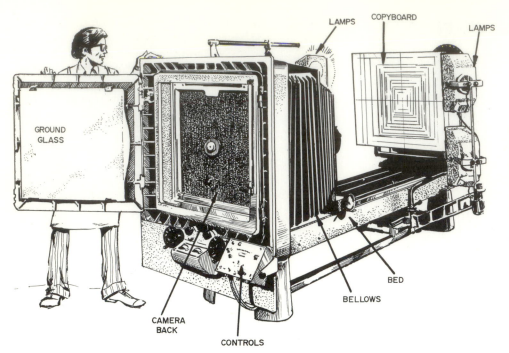

FIGURE 274 A process camera which photographs the image of the copy to the exact size of the final reproduction

A sheet of high-contrast photographic film is fitted into the camera back, held in place by a vacuum, just as the copy being photographed is held flat against the copyboard by a vacuum. The photographer focuses the image of the copy onto the large ground glass in the back. By adjusting the position of the bellows along the camera bed, one can enlarge or reduce the image on the film to the exact size of the final reproduction.

line cut. The mechanical used has *line copy,* copy composed of solid blacks and whites without true grays. Fig. 272 is also an example of line copy.

Photoengravings are used in letterpress printing, which is "relief printing," the kind most familiar to the average person. In relief printing the image is printed when ink is carried to the paper from a raised surface. Fingerprinting is a common example. Stamp pads ink the relief surface of a rubber stamp. Typewriter keys strike the ribbon with the relief surface of the key. Printer's type is another example, and, indeed, is the source of the name *letterpress.* In times past the illustrations for letterpress were printed from wooden blocks which had been carved so that only portions remained in relief. We have seen two such illustrations already, both by Dürer (see Figs. 275 and 276). The procedure for carving and printing such a block is described in Fig. 277. Blocks of this type may be printed by hand, but if they are to appear alongside type, the usual tech-

FIGURE 275
ALBRECHT DURER. *Demonstration of Perspective, Draftsman Drawing a Lute.* From the artist's treatise on geometry. 1525. Woodcut, 5¼ × 7⅓". Kupferstichkabinett, West Berlin

FIGURE 276 ALBRECHT DURER. *The Four Horsemen of the Apocalypse.* c. 1598. Woodcut, approximately 15¼ × 11″. Metropolitan Museum of Art, New York. Gift of Junius S. Morgan, 1919

nique is to use an absolutely true block which is locked into the *chase* of a press.

There are three basic types of letterpresses. The *platen press* has two flat surfaces—a bed, which carries the plate and/or type, and the platen, which supports the paper. Dürer's woodblocks would have been printed on a flat-bed platen press of the sort Johann Gutenberg (c. 1397–1468), reputed inventor of movable type, used. The kind of platen press still in use today—used for short runs of stationery, cards, and the like—is a clamshell platen press. It is the opening and closing mechanism you have probably seen in newspaper offices in Western movies. A *flat-bed cylinder press* prints much more rapidly; it has a large revolving cylinder instead of a platen. The most advanced form of letterpress is the *rotary press,* which uses a plate cylinder and a printing cylinder as well. Unlike platen and flat-bed cylinder presses, which print on separate sheets of paper, the rotary press can be *web-fed* from a continuous roll of paper that is cut into sheets after printing. It is highly accurate, prints at much greater speed than other letterpresses, and is used for jobs requiring a great number of very sharp, high-quality copies. Today most letterpress printing in any quantity is done on rotary presses, but because the mechanics of letterpress printing are most easily understood in terms of the flat-bed cylinder press, we have chosen it for our model. In Fig. 278 you see the functioning elements stripped of the protective coverings and machinery that makes it possible for them to work.

(A) This is the image the artist wishes to print. Notice that the rectangle is to the left of the circle. (B). When the printmaker draws the design onto the plank, the image will be reversed, because all relief prints come out backward. (C) The printmaker then chisels away everything that is to be white when the block is printed. (D) With a soft rubber or gelatin roller called a *brayer,* the block is inked with oil-base printer's ink. Soft thin rice paper (E) is placed on top of the block and rubbed with a wooden spoon or similar implement so that the moist ink is transmitted from the relief surface to the paper. (F) An impression is pulled. The block is reinked for the next impression, the rubbing repeated, and so on, until an edition is complete.

FIGURE 277 Preparing and printing a woodblock

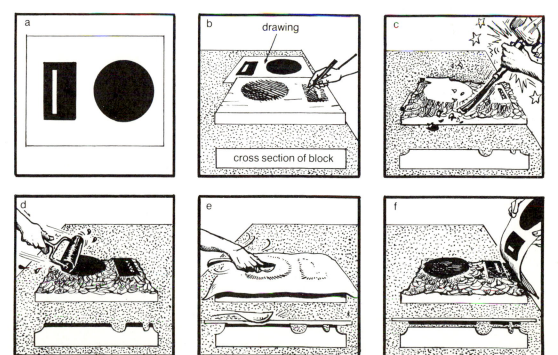

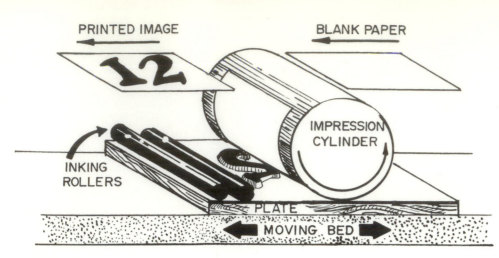

FIGURE 278 The flat-bed cylinder press

A sheet of blank paper is seized by a *gripper* device on the impression cylinder and is taken around to be pressed against the plate, which rests face up on the bed of the press. The bed itself moves back and forth in such a way that in the time it takes a sheet of paper to whip around the cylinder and be released the bed has passed the plate beneath ink-covered rollers that transmit the sticky printer's ink to the raised surfaces on the plate. As the bed carries the inked plate forward again, the impression cylinder turns in answering motion and brings the paper into contact with the plate, impressing the image onto it. The cylinder lifts slightly to free the bed, carries the paper clear around to release it, and grips a new sheet. And so it goes, for as many impressions as the job requires.

photomechanical reproduction processes: offset lithography

At one time practically all commercial printing was done by letterpress, and much of it still is, especially on small production runs. However, the most flexible and convenient of the printing techniques is of much more recent origin. In fact, it is the way the book you're reading was printed. It is *offset lithography*.

Lithography means ''stone drawing'' and derives from the origins of the technique in the fine arts in which the printing surface was originally (and usually is still) the smooth face of a block of Bavarian limestone. Unlike all other printmaking media, which are mechanical in nature, lithography is chemical. It depends upon the fact that oil and water will not mix. Honoré Daumier's political comment on a momentary respite in political tension (Fig. 279) is a handmade lithograph. That is, Daumier actually drew the image in reverse onto a slab of limestone, which was then processed and used to print the image appearing in the left-wing gazette, *Charivari*. The process of creating and printing a lithograph is explained in Fig. 280. (Incidentally, the crayon texture has been imitated here the same way political cartoonists imitate it today—by use of a black crayon on a pebble-grained paper called ''coquille board.'')

Offset lithography is the modern, commercial version of lithography, employing plates of zinc or aluminum instead of stone blocks. Negatives are prepared exactly as they are for letterpress and the making of the plates is very similar; it is the printing pro-

FIGURE 279 HONORE DAUMIER. ''Well for once it looks as if Papa Mars does not intend to cut the grass from under my feet.'' 1866. Lithograph, 8 × 10½″. Private collection

Daumier's cartoons in the left-wing Parisian press of the nineteenth century were genuine handmade lithographs printed from limestone slabs. The reference in this one is to an unexpected reprieve from the threat of war during a period of great political tension.

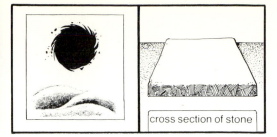 

FIGURE 280 Preparing and printing a lithograph

(A) The image to be produced is like the one on the left. It will be drawn on a slab of Bavarian limestone. (B) Solid areas are painted in or penned on with a greasy substance called "tusche," which behaves more or less like ink.

(C) For graduated areas, tusche in the form of crayons is used. (D) The stone is then "etched" with a solution of water, gum arabic, and a few drops of nitric acid. Where the gum comes in contact with tusche, it makes the tusche insoluble in water. The acid bites the limestone just enough to make it slightly more porous than before and therefore more capable of absorbing water. (E) Then the artist washes out the image with water and turpentine. At this point the stone is ready for printing. This, again, depends upon the antipathy of greasy areas and water.

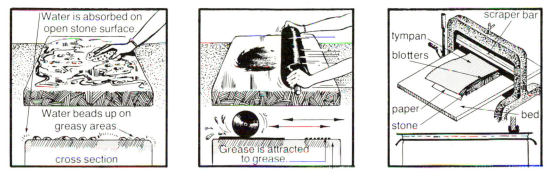

(F) The stone is wet down with water, which is absorbed by the open stone but beads up on the areas where tusche had been applied. (G) With a previously inked roller the size of an extremely large rolling pin but covered with rubber or soft leather, a few passes are made over the wet stone. The greasy ink is repelled by the open, wet areas, but in greasy areas the roller pushes the beads of moisture out of the way and the oily ink is attracted to the oily surfaces. As if by magic, the original drawing reappears on the stone. (H) A damp sheet of printing paper is laid over the stone, which has all this while been resting on a press bed. The paper is topped with a couple of blotters and a piece of red fiberboard called a tympan.

(I) The press has a scraper bar which has been lubricated with mutton tallow. It is clamped down onto the similarly well-lubricated tympan and presses the paper hard upon the stone. (J) The printer cranks the press handle, driving the movable bed along under the scraper bar. The stone, of course, is carried along too and the pressure exerted by the bar squeezes the image from the stone onto the paper. (K) The impression is pulled from the stone, the stone is rewet and reinked, another sheet of paper is applied, and the whole business is run back through in the other direction. This process is continued until the edition is completed. Once an edition is finished, the stone is ground down to remove the image entirely and another lithograph may then be drawn upon the same stone.

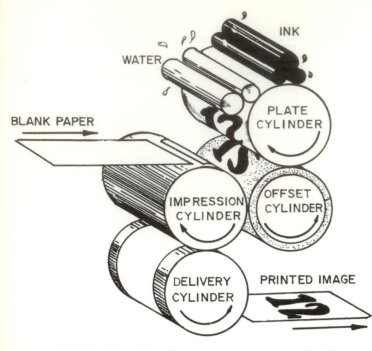

WATER INK BLANK PAPER PLATE CYLINDER IMPRESSION CYLINDER OFFSET CYLINDER DELIVERY CYLINDER PRINTED IMAGE

FIGURE 281 Offset lithography, a modern and flexible printing technique

Once an image has been fixed into the lithographic plate, the thin plate is wrapped around a drum on the offset press. When the press is running, the plate cylinder revolves and, as it does, it is contacted first by a set of rollers that wet it with water and then by a set of rollers that apply printer's ink. The water beads up on the grease-receptive areas but is absorbed into the others. Since the bare metal areas are wet, the greasy ink will not adhere to them, but in greasy areas the ink rollers push the drops of water out of the way and the ink is attracted to the oil surface. Next, the inked plate contacts another drum, this one wrapped in a rubber blanket, and the inky image is "offset" from the plate onto the blanket. Blank paper is brought against the blanket by an impression cylinder, printed, and then removed and deposited by the delivery cylinder.

cess that differs. The plate is a very thin sheet of zinc or aluminum (about as thick as a coffee can) so treated that after it has been exposed to light through a negative, as in photoengraving, and lightly etched, the zones where light was excluded (black in the negative but white on the mechanical) are receptive to water and the places where light hardened the surface are receptive only to grease or oil. Printing, then, depends upon the fact that printers' inks are oil-based greasy substances having an antipathy to water (see Fig. 281).

Offset lithography permits finer printing on rougher and poorer grades of paper than letterpress and has such other advantages as higher speed and longer-lasting plates. Practically all magazines, brochures, and illustrated books are printed by this method. The images are not quite as sharp and clean as good letterpress work, however.

other photomechanical processes

There are other major modes of photomechanical printing in the graphic arts. Currency, stock certificates, and many fine books are printed by *gravure,* which is capable of tremendous subtlety and refinement and can also do very high-speed runs of massive numbers of impressions. It is, however, an expensive and quite complicated process and is not too often encountered by graphic artists. Similarly, the *collotype* process, which prints from gelatine plates, is useful only for the most expensive and exquisitely complicated kinds of reproduction. Bus placards, theatre posters, and similar advertisements are often printed by *silk screen,* a stencil process which can be used to do fine arts prints called "serigraphs" as well as ordinary commercial work. It is an extremely flexible process and, although it is peripheral to our interests here, it does afford opportunities to designers of everything from posters to wallpaper and drapery fabric.

tints, halftones, and combination plates

All of the noncolor reproductions in this book were done by a process that reduced everything to black and white without grays. That includes the reproduction of the photograph of the mechanical in Fig. 270 as well as the rendering in Fig. 267. The original photograph and the painted design are examples of *continuous tone copy,* as opposed to the line copy used in our earlier example. That is, the photograph and the painting actually do contain a full range of grays between black and white, unlike line copy, which merely may appear to contain grays. Line copy is used to create a line cut whereas continuous-tone copy ends up in what is called a *halftone.* What a halftone does is reduce all of those grays into minute dots so that they can be printed in exactly the way a line cut can. The sole difference is the way in which the copy is exposed to the process camera.

There are several ways of making halftones, but the principle is always essentially the same. Here is one method: A halftone *screen* is placed in the process camera between the lens and the film (see Fig. 282). A halftone screen is a sheet of glass or plastic engraved with very fine diagonal lines that have been filled with black pigment. Because the film used in reproduction photography is incapable of discriminating anything except black or white, it "sees" the copy

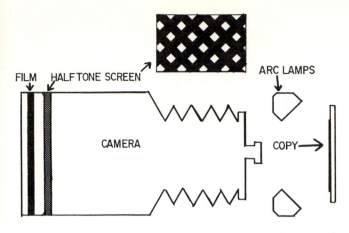

FIGURE 282 The diagram represents one system for producing halftone negatives

in relief; there are many dots close together where the image is to be black and very few where it is to be highlighted. If you look at a newspaper, you will see that reproductions of photographs follow this pattern. Newspapers use very coarse, 65-line halftone screens. Screens of 100 lines are standard for the kind of paper you are reading from right now. Finer screens (120 to 150 lines) are suitable only for *coated paper,* paper whose surface has been covered with pigment and adhesive to improve the printing surface. (Usually, these are glossy stocks but they may be dull surfaced as well.)

If you make a halftone version of a solid black area, it will produce what is known in the printing trade as a *tint* (Fig. 284), that is, a flat tone consisting of uniform dots equidistant from one another. Such tints are specified as percentages of solid black. Thus, a 100 percent tint contains no white at all; a 50 percent tint is exactly halfway between white and black; a 10 percent tint is very faint gray. The determining factor is the size of the evenly distributed dots. The 10 percent tint contains tiny dots, the 50 percent tint larger ones. Any line copy may be reproduced as a tint in any percentage of black desired (Fig. 285). In photoengraving, the tints are usually called *benday,* after Benjamin Day (1838–1916), who invented them as prefabricated patterns consisting of different sized dots and lines in densities varying from extremely light to solid black. There are also white versions to use against black backgrounds, graduated patterns, and exotic textures. Originally, these dots and lines had to be incorporated into the art work by printers during the platemaking process, but today they are available to artists and designers for direct use in the

in terms of those extremes. Without a screen in the camera the image would be reduced down to blotches, like the image in Fig. 283A. When a screen of 100 lines has been interposed, these gross patches are broken up into tiny points of light and dark. As it happens, optical law dictates that a light area the same size as a relatively darker area will look relatively larger, to a camera as well as to the human eye. Therefore, the high-contrast film sees white areas of copy as made of large dots, gray areas of slightly smaller dots, darker gray areas of still smaller ones, very dark areas of tiny dots, and black areas as having almost no dots at all, producing a negative like the one in Fig. 283B. A plate made from that negative, in the same way a line cut is made, prints the kind of image seen in Fig. 283C. Enlarged to the same degree as Fig. 283B, it will have a lot of little dots standing up

FIGURE 283 A: Continuous tone copy shot without a screen; B: Enlargement of a section of a negative made with halftone screen, ten times; C: Continuous tone copy shot with 100-line screen and reproduced; D: Enlargement of section of halftone reproduction by 10 times

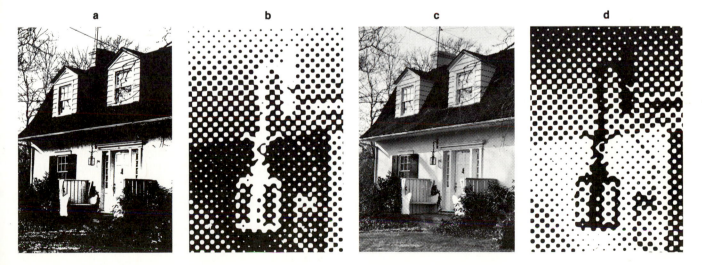

a   b   c   d

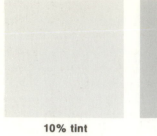   

| 10% tint | 30% tint | 50% tint | 60% tint | 90% tint |

FIGURE 284 Halftone tints

a

FIGURE 286 Line drawing tinted with benday patterns on shading sheets

b

FIGURE 285 A: Line copy reproduced as rendered, in solid black; B: The same copy photographed through a halftone screen to produce 40% tint

form of *shading sheets*—they have been printed onto clear plastic film backed with a pressure-sensitive adhesive. The gray areas in Fig. 286 were achieved by application of pieces cut from such sheets.

Fig. 286 combines halftone tints with line drawing but it is not what is called a *combination plate* because the shading film is itself a form of line copy. Fig. 287, however, combines the elements in A (a line

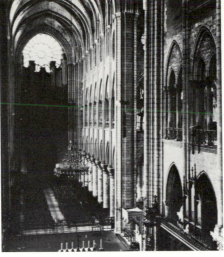

a

b

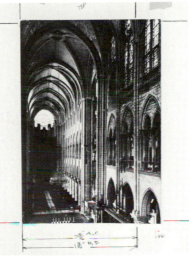

c

FIGURE 287 A combination plate in: (A) its separate elements; (B) reproduced form; and as a (C) mechanical. The elements were combined in negative form (D) by simply overlapping them (E).

Since the areas that were black in the mechanical are transparent in the negatives, while areas originally white are opaque in negatives, the plate is exposed to light only where it is to be black. Opaque areas of the halftone negative block out clear areas of the "window." The result is (B) a combination plate. Photograph of Notre Dame, Paris, Courtesy French Government Tourist Office, New York.

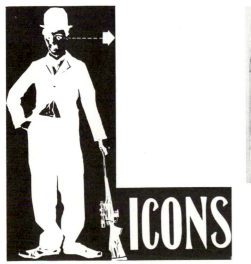

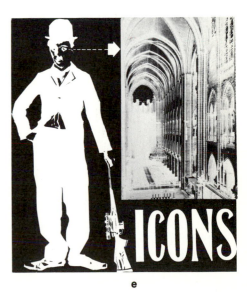

d

e

drawing and continuous tone photograph) B, and it is therefore a combination. The mechanical for this simple job is shown in Fig. 287C. As you can see, the halftone was reduced in size (having been scaled and cropped as demonstrated in Fig. 109) and fitted into the space indicated by the black square. Most people are puzzled by the fact that the black square does not obliterate the halftone picture. You must realize that all plates are made from negatives which are opaque and transparent. Figures 287D, 287E, and 287F show that the black square is really a window for the halftone negative. An outline of the area, drawn in red, will accomplish the same thing; such an outline is known as a *holding line* or a *key line*.

FIGURE 288 The steps in producing a dropout combination plate.

A: The elements of the mechanical.

THOMAS COUTURE. *The Enrollment of the Volunteers of 1792.* 1848. Oil on canvas, 23 × 40″. Museum of Fine Arts, Springfield, Massachusetts

B: The negatives of the elements (For line copy, the negative is "reversed," that is, made positive)

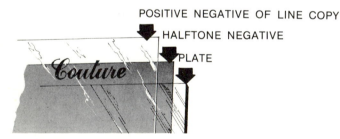

C: Exposing the plate

D: The result of combining the elements

The same principle of overlapped negative transparencies enables designers to run white line elements through screened areas. Frequently, what looks most difficult in graphic art is really very easy to accomplish. Let us assume that you wish to run some white lettering and drawing across a halftone. It would certainly be possible to paint them directly onto the photograph although to do so would be tedious and time-consuming. Too, it would result in lettering that was ragged-looking because the halftone screen used to shoot the platemaker's negative would impose its dot pattern on the edges of each letter. Doing the lettering so that it is sharp-edged line copy not only would look better but would be easier. It requires nothing more than the overlapping of the negative for the halftone with the one carrying the line copy to produce what is known as a *dropout* (Fig. 288). If the designer wishes the lettering to be black instead of white, he or she need only write the word *surprint* instead of *dropout*. This process, known as *double-burning,* is described in Fig. 289. Combinations of dropouts and surprints are common in graphic production. Also, in what amounts to a one-step dropout effect, platemakers can easily produce a

FIGURE 289 The steps in producing a surprint combination plate

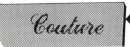

A: The elements of the mechanical are identical to those of the dropout except for the written instruction

B: The negatives of the elements

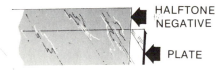

C: Exposing the plate twice, separately for halftone and line copy, in what is called a "Double-burn"

D: The result of combining the elements

a

b

FIGURE 290 Design for stencil lettering
A: Line copy reproduced as a conventional positive
B: The same copy "reversed"

negative version of positive copy (Fig. 290). This is called a *reverse* and requires from the designer only a note to the printer saying: "reverse negative."

color printing

In Chapter 6 we spoke of the tiny phosphor dots of the television set which, upon being illuminated, produce the spectral hues in varying intensities, depending upon the combinations of dots stimulated and the strength of the light beam striking them. The tiny phosphor dots of the TV screen are additive versions of the kinds of images reproduced by a subtractive process for the colorplates in this book. All color reproductions are made up of tiny halftone dots which separate the full color of a painting, for example, down into the three process colors, magenta, cyan, and yellow. The technique of doing this is called "process color separation" and it is done by platemakers in printing plants from original art or from high-quality color transparencies similar to 35-mm color slides but five or ten times that size. (The increase in scale leads to a comparable increase in optical quality.) Process separation involves very close tolerances and a real understanding of it requires a more than rudimentary grasp of color theory. Colorplate 22, however, provides a visualization of the essential idea.

Any of the standard painting media can be used to produce art work for color process separation, but

it is wise to avoid certain kinds of color combinations. In particular, intensely red forms painted into black areas are difficult for the camera to discriminate (since, regardless of the filter color, the film itself is black and clear) and will entail special work by the platemakers, thereby increasing costs.

Graphic designers may make bold use of process color in their work but they do not often have direct association with the actual making of transparencies or separations. Correcting reproduction plates for color accuracy according to the transparencies is highly specialized and far beyond the scope of a general text. The Bibliography, however, has books concerned with just such matters.

Normally, the graphic designer has direct contact with color separation in the studio by way of what is called "mechanical color separation," which uses color overlays of magenta, cyan, and yellow along with black to produce "full color" effects. One most commonly encounters such effects in the comic strips. Comic books and the Sunday funnies are usually printed in four colors from mechanical separations done in the studio. "Separation" can be taken quite literally; it is any process whereby colors such as green, violet, brown, orange, pink, tan, or whatever are broken down into their four process components. Process separation is an entirely optical procedure; mechanical separations (Colorplate 23) are done by hand.

Obviously, any number of graphic artists, not just those who do cartoon strips, make use of the

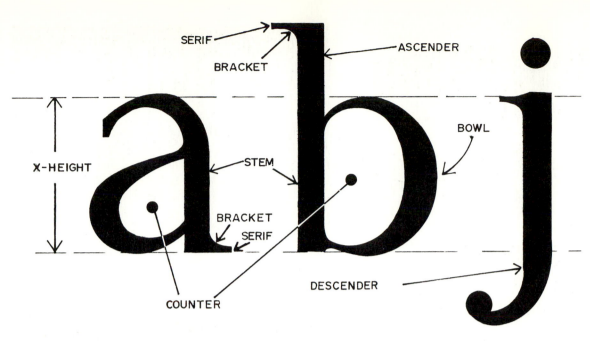

FIGURE 291 Lettering terminology

The bowls and stems of the lower-case letters establish the "x-height" of a given alphabetic style in any specific size. When the stem is extended above the x-height, it is called an *ascender*. When it drops below, it is called a *descender*. The open space in a letter is termed a *counter*. The little prominences on the ends of these characters are called *serifs*, and the rounded junction of the serif with the stem is referred to as a *"bracket."*

same techniques. Indeed, virtually all graphic design entails at least some mechanical color separation work. Most often one is combining nonprocess hues for decorative impact (as in Colorplate 24) rather than in imitation of natural effects. The techniques used, however, are essentially the same.

lettering

Although the fundamentals of lettering are quite simple, the fine points of typography, hand lettering, and calligraphy are too many and diverse for us to cover in an introductory text. Essentially, all of the characters in the Roman alphabet—the alphabet for Latin, used also by the English-speaking peoples—are made up of elements of triangles, squares, and circles. Numerous refinements have modified these simple shapes for the sake of legibility and attractiveness. There is a special terminology associated with lettering forms. Figure 291 is a diagram exemplifying some of the principal letter elements and the terms assigned to them. Of the *characters* (that is, the letters, numerals, and punctuation marks), those called "letters" fall into two categories. The ones in Fig. 291 are *lower case*. Laypeople usually call them "small letters." Printers and designers refer to what laypeople call "capitals" as *upper case*. (Strictly speak-

ing, a capital letter is a letter at the head of a word.) These two terms derive from the positions of the boxes holding type in hand typesetting. Not all letter styles contain all of the elements described in Fig. 291. Obviously, upper-case characters do not have ascenders or descenders. Also, in certain styles, there are serifs without brackets. For instance, the Clarendon typeface (Fig. 292) has boldly bracketed serifs; Egyptian (Fig. 293) is similar to Clarendon but is without the brackets. There is also a family of letter styles that lack serifs altogether. They are called *sans serif* (that is, without serifs). Helvetica (Fig. 294) is a noteworthy example.

FIGURE 292 Clarendon typeface

FIGURE 293 Egyptian typeface

FIGURE 294 Helvetica typeface

Helvetica

Normally, alphabets are divided into Roman styles (having thick and thin elements and usually serifs), *sans serif* styles (without serifs and frequently of virtually uniform thickness), and *display* styles (Fig. 295). Within each category there are *italic* versions, in which the characters slope at a uniform angle (Fig. 296). Also, lettering can be *condensed* (Fig. 297A), *expanded* (Fig. 297B), made *light* (Fig. 297C), *bold* (Fig. 297D), or *extrabold* (Fig. 297E).

In Figs. 298, 299, and 300 three lettering styles are worked out within grids so that the relative proportions of the elements and the characters themselves will be apparent. The characters used here are the simplest and most commonly encountered versions. It is possible to expand or condense the letters merely by applying anamorphic principles (as in Exercise 2–3), and one can reduce or enlarge them simply by contracting or magnifying the grid (Fig. 301).

FIGURE 295   A miscellany of display typefaces

Helvetica

FIGURE 296   Italics

Helvetica

FIGURE 297   A: Condensed lettering

Helvetica

B: Expanded lettering

Helvetica

C: Light lettering

Helvetica

D: Bold lettering

Helvetica

E: Extrabold lettering

FIGURE 298   Roman style alphabet

FIGURE 299   Sans-serif style alphabet

FIGURE 300 Block condensed style alphabet

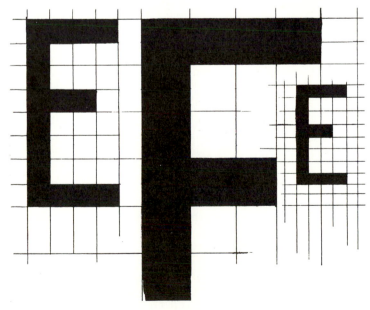

FIGURE 301 Enlargement and reduction of grid and letter

197

SPACE

a

SPACE

b

FIGURE 302 A: Mechanically spaced lettering; B: Visually spaced lettering

Something you cannot do is include serifs on characters from a sans serif alphabet. Lettering styles are either sans serif or not, and an upper-case "I" does not have serifs in a sans serif alphabet.

A vital consideration in graphic design is the spacing of the characters, words, and lines of lettering. The first rule—a hard and fast one—is that one *never* spaces letters mechanically except on a machine where avoidance of regularity is impossible. Mechanically spaced characters look uneven (Fig. 302A) because, while the distances are equal, the areas are not. Designers therefore strive to make the spaces between the letters look uniform in total area (Fig. 302B), using visual judgment rather than measurement. One caution: in dealing with open-face letters, such as upper-case Cs or Es, you don't "count" the whole space within the opening, only about one sixth of it.

The second rule regarding spacing is that interline spacing (the distance between one line of words and another) should be consistent. The lines should

FIGURE 303 A: Pointed and rounded forms improperly drawn *to* guidelines

B: Pointed and rounded forms properly drawn *just over* guidelines

be parallel and far enough apart to avoid collisions between ascenders and descenders but not so far apart that they seem isolated from each other.

Third, it is important to bring the points of letters like A, V, N, M, and W slightly over the guidelines. That is also true of the curves at the tops and bottoms of rounded letters C, G, O, Q, S, and U. If these letters are not imperceptibly larger than their fellows, they will look smaller (see Fig. 303A, and B). It's an optical illusion. Of course, sometimes these letters don't have a shape requiring such treatment. For example, the condensed sans serif alphabet in Figure 300 has no truly pointed or rounded characters.

Since few of us are skilled in the creation of letterforms, it is common among designers to use prefabricated lettering in the form of *transfer type*. Transfer type is basically of two kinds—cut-out and rub-off.

With cut-out transfer type the elements of a font of type are printed on a sheet of transparent film the back of which has been coated with a transparent waxy adhesive. The section containing a given character is cut out with a razor blade or frisket knife and placed on the art work. Rubbing with a burnisher of some sort will adhere the film, along with its character, to the art. Polished steel burnishers, metal spoons, rounded objects of hard plastic, pencil-sharpened dowel sticks, and numerous other things can be used to burnish the film. A commonly available brand of cut-out transfer type is Artype.

Rub-off transfer type, of which the best known line is Letraset, is similar to cut-out except that both the characters and the adhesive are printed onto the back of the plastic sheet. The artist simply positions a character over the art work and rubs it with a burnisher; since the wax adheres to the paper, it carries the character with it and, when the film is pulled away, the letter remains on the art work.

Reproduced, the cut-out and rub-off transfer types are scarcely distinguishable. Fig. 304A is a photograph of Artype as it appears on the work; Fig. 304B is the same thing printed. Fig. 304C is Letraset on art work and 304D the printed version. As to advantages and disadvantages, the cut-out types are stronger and will retain sharper detail in reproduction. The rub-off types are easier to use and look cleaner and neater on the original art work. (Cut-out type need not be quite so coarse in effect as it is in Fig. 304A, though. We were exaggerating for the photograph. But rub-off types look just as if they had been printed on the paper.)

The ease and economy of lettering with transfer type have overcome nearly all objections to it. One objection is that the character of the lettering does not change in weight when it is reduced and therefore it

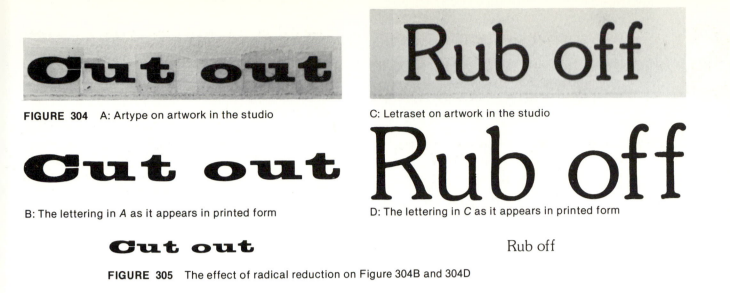

FIGURE 304   A: Artype on artwork in the studio

C: Letraset on artwork in the studio

B: The lettering in *A* as it appears in printed form

D: The lettering in *C* as it appears in printed form

Cut out   Rub off

FIGURE 305   The effect of radical reduction on Figure 304B and 304D

looks weaker. Virtually all graphic designs intended for reproduction are done somewhat larger than they appear in print, usually at least 25 percent and often two or three times larger. When a given typeface is reduced in size, it tends to weaken. Fig. 305 shows how the lettering in Fig. 304 looks when it has been markedly reduced. The counters seem to be larger and the strokes weaker. For this reason, people who design lettering compensate by making the small sizes of type bolder than the larger ones (and vice versa, since enlargement of a small character makes it look overweight). This objection is valid, but the alternatives to transfer type are so expensive or troublesome that only the most fastidious designers

will hesitate to make use of it. Still, one ought to try to select type sizes that will suffer as little as possible by reduction.

When lettering is used in designing, an excessive variety of forms ought to be avoided. What is excessive? How many typefaces can one have in a single design? What is the maximum number of different kinds of drawings that can be installed in a single design? These things are not easy to stipulate, even as rules of thumb, because what is excessive in most cases may be desirable in some specific instance. Fig. 306 is a case in point. In this presentation an overabundance of letter styles gives an impression of endless alternatives. On the other hand, it has no

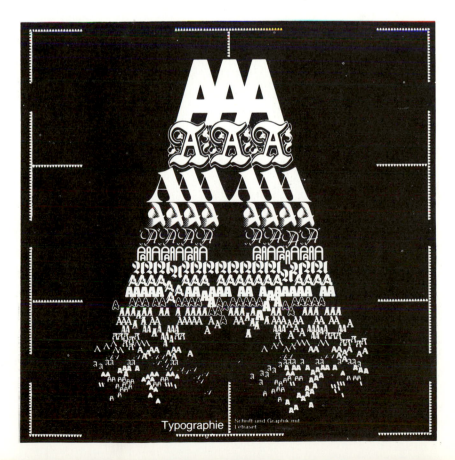

FIGURE 306   A design featuring a multitude of typefaces. Christof Gassner, Cover of catalogue for exhibition "Typography and Graphics with Letraset" in Germany

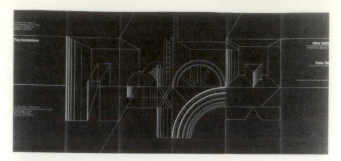

FIGURE 307 Designs featuring a lot of open space
FIGURE 307A ROBERT STEARNS, Art Director. STAN BROD, Designer. Exhibition Catalogue, Contemporary Arts Center, Cincinnati, Ohio

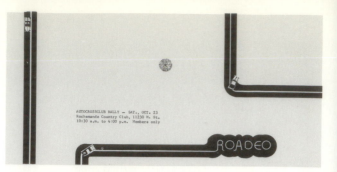

307C "RI". Special mailer for a country club activities group. 1981.

307B ARRESSICO. Record jacket for "Tear Zero" band. 1982.

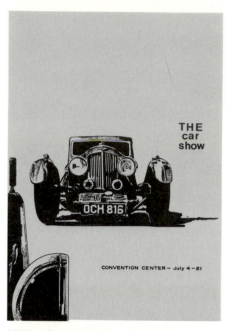

307D JOHN ADKINS RICHARDSON. Antique automobile show poster. 1980

307E ARRESSICO. First statement of concern to client whose account is in arrears. Cover. 1982

307F GLENDA WASHINGTON. Cover of public service brochure explaining why lawns should be watered in the evening. 1979

a   b   c

FIGURE 308 A record jacket incorporating the word "Jazz" in different ways

message other than that to convey. *Generally speaking, it is unwise to use more than three styles of lettering in a single design. Two is safer yet. When in doubt, "sell" blank, white space* (Fig. 307).

One last caution. Since the lettering is carrying information, if not the main weight of the message, it should be easily distinguished from its surroundings. Therefore, a degree of contrast which in a work of fine art would seem unharmonious is desirable in some graphic design. Thus, if we were to use something like Davis's *Colonial Cubism* (Colorplate 4) for a jazz record jacket, it would be wise to superimpose against it lettering forms of a contrasting color and character. Compare, for example, the styles in Figure 308. In Fig. 308A the lettering is so integrated into the design as to be invisible. The words in Fig. 308B stand out all right, but they are utterly out of character. Of the three, Fig. 308C is the best solu-

tion; the lettering here is legible, attention-getting, and also suitable for the context in which it appears. Ordinarily, this kind of thing would be worked out in preliminary studies known as *thumbnails, layouts,* and *comprehensives.* We have already considered thumbnail sketches in connection with illustration (see Exercise 4–4 in Chapter 4). They are tiny, very rough notations of ideas (Fig. 309). A layout is a preliminary plan for the design in which all of the basic elements are shown in correct positions and proportions (Fig. 310). A comprehensive (Fig. 311) is commonly referred to as a "comp." It is a rendering intended to give viewers a very clear impression of what the printed version of the design will be like. The designs in Fig. 311 were painted in designer colors to give a correct presentation of the ultimate designs.

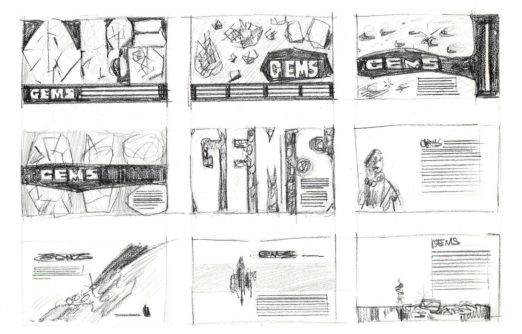

FIGURE 309 Thumbnail sketches for a graphic design

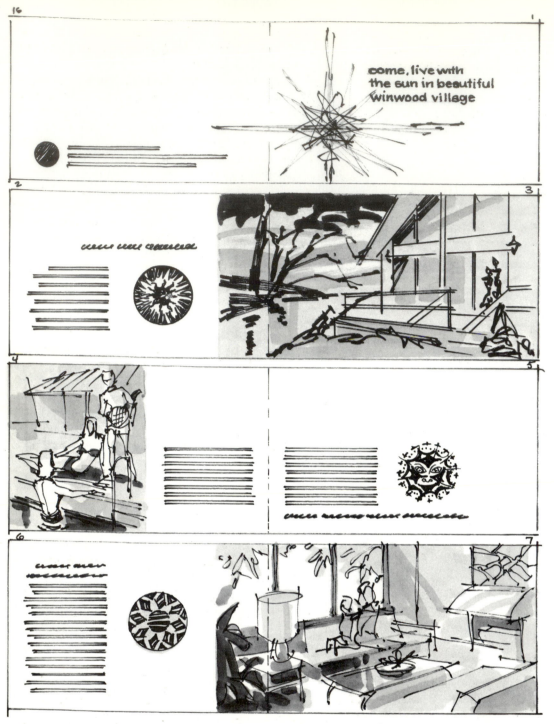

FIGURE 310 Layouts for a brochure advertising a condominium. From Bud Donahue, *The Language of Layout,* p. 168 (Prentice-Hall, Inc., 1978). Reprinted by permission of Prentice-Hall, Inc.

FIGURE 311 Comprehensives or "comps." In effect, a depiction of what the reproduced design will look like in proportion, color, and character. These were student designs for record albums, c. 1960.

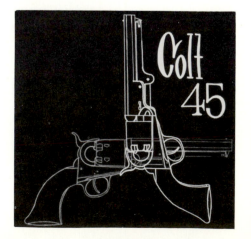

Obviously, in designing a piece of graphic art one begins with thumbnails, proceeds to layouts, then does a few comprehensives, and finally works up the mechanical. Still, it is helpful to have a few preliminary notions of how the design might possibly be organized before one begins doodling on a note pad.

layout organization

Fig. 312 shows two pages of a pamphlet upon which a grid has been imposed. The shaded cells provide us with a simple six-unit pattern, itself a sort of grid. The pattern could have been made up of two units (three cells wide and eleven long) or four (three wide, five long) or three (seven wide, three high) or eight (three wide, two high) or some other configuration, but the one given is the most obviously useful arrangement.

When one is dealing with pages in a folder, magazine, or book, a particular nomenclature is employed and we have spelled it out here. Right-hand pages are called *recto* (Latin for ''right'') and left-hand pages *verso* (being the *reverse*, or back, of the recto side). *Margins* are external borders and *gutters* the internal ones, next to the centerfold. Divisions within a page are called *alleys,* whether they are horizontal or vertical. Usually, the lower margin is

slightly larger than the others and it frequently carries what is called the *folio* (page number) and perhaps the volume and date.[1]

At least some of the material in any page layout will coincide with the margins, gutters, and alleys. In Fig. 313A illustrations (denoted by black areas) fill the left-hand columns and typeset captions (indicated by parallel lines) appear on the right. In Fig. 313B the illustrations have been staggered with the boxes of type. In this case the recto page mirrors the verso page; it could, instead, have duplicated it just as the recto page in 313A might have mirrored the verso page. Both of those layouts are common and useful. Neither is imaginative but either would be suitable for publications that list a number of things of approximately equal status or of a certain character. Class yearbooks, realty listings, tourism brochures, and so on are all examples of material employing layouts of this kind.

Fig. 313C varies the use of the grid, for here it has been disrupted by illustrations crossing alleys in the verso page and, on the recto side, by the deletion of a unit from the top left corner. Presumably, the printed message on the right refers to the pictures collectively.

Margins or gutters are no more sacrosanct than alleys, and in Fig. 313D illustrations have been allowed to run off of the page. This device is called

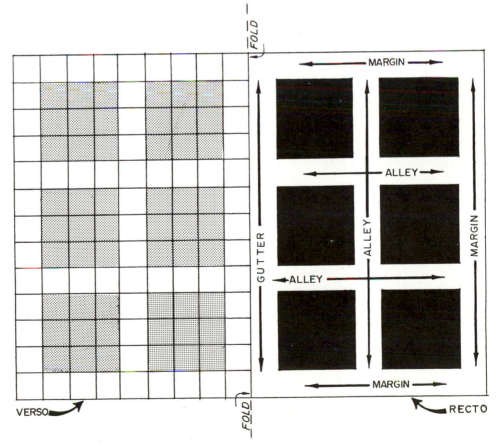

FIGURE 312 A two-page layout showing printing nomenclature.

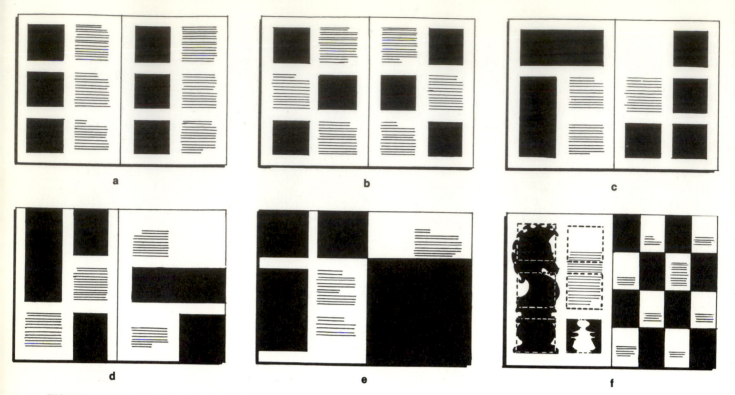

a b c d e f

FIGURE 313 The grid as the basis of page layouts

FIGURE 314 A Nieman-Marcus advertising brochure featuring a very free ink drawing against a grid of tints and halftones

bleeding. When illustrations are permitted to run all the way over to the fold, they are said to *butt the gutter* (of the other page, that is). In Fig. 313E a very large recto illustration crosses alleys, butts the gutter, and bleeds off of the bottom and right edges of the page. Too, the lettering above the large illustration crosses an alley; its left edge would be aligned with the right-hand edge of the top left box unit in our grid.

Fig. 313F shows how a grid (indicated verso with dotted lines) can deal with an irregular, silhouetted shape as well as with rectangular units. Here, the letter has been made to cross over the top horizontal alley and begin low in the upper right module. The first line of this lettering is on the same level as the uppermost horizontal division of the recto page, helping to bring a grid of a different proportion into harmony with the one we have been using up to now.

Obviously, these are but a few of many possibilities. For instance, Fig. 313C could have had type where the illustrations are and illustrations in place of the type. Nor did every illustration in Figs. 313D and 313E have to be bled. In Fig. 313F type might have filled one or more of the boxes of the checkerboard pattern.

The grid system of laying out pages is fre-

quently encountered, so frequently that it must be considered something of a cliché. But, like all clichés, it has become one only because it is so useful, so universally applicable, so comfortingly reliable, and so effective, even though it is rarely judged imaginative. Disruption of the grid's purity (Fig. 314) can produce an impression of spontaneity, but even this is not a very lasting impression. Something more decisive may be called for.

A collapse of formality that does not convey the disharmony of utter chaos is evident in Fig. 315. There are exactly the same number of black and gray rectangles in both Figs. 315A and 315B but the grid arrangement in A is absolutely certain and predictable. Of the arrangement in B it can be said only that there is, in its randomness, a certain probability of satisfying the viewer. Creating layouts based upon the latter kind of intuitive orderliness is far more difficult than basing them upon grids, but there is something peculiarly appealing to modern eyes in the informal yet coherent aesthetic of the arrangement in Fig. 315B. Fig. 316 contains several designs that make a similar kind of appeal. Of course, they are not adaptable to innumerable situations in the way the grid is, but it is just their individual uniqueness that makes them strong.

FIGURE 315 Different kinds of order achieved with identical elements. The amount of space in A and in B is identical and there are the same number of black and gray rectangles, but the principles of arrangement are entirely different.

a

b

a

b

FIGURE 316 Designs based on informal, intuitive arrangements.
A: Booklet cover. Arressico, 1982
B: Nipon Design Center
C: Katherine McCoy, Design Director, James Houff, Designer: Exhibition Catalogue, Cranbrook Art Museum
D: Bruno Oldani, Record jacket for Nordisc

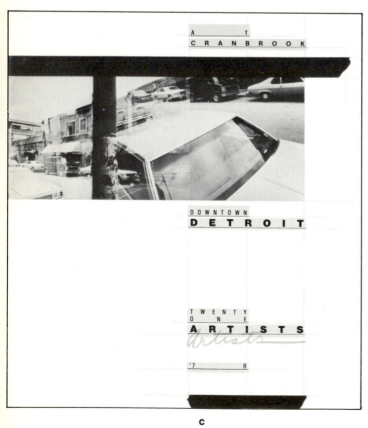

c

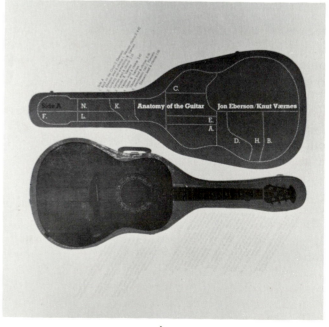

d

EXERCISE 9–1: Transfer Type

Materials: White drawing paper
Transfer type
Tools: Single-edge razor blade or frisket knife
Burnisher, teaspoon, or similar implement

With transfer type alone or in combination with other elements, do a typographic design in which the arrangement or form of the characters is itself used to reinforce the message being communicated. A few examples appear in Fig. 317.

EXERCISE 9–2: Shading Film

Materials: White drawing paper
Shading film tints
India ink
Tools: Single-edge razor blade, frisket knife, or pin vise
Burnisher, teaspoon, or similar implement
Drawing instruments

Create a design for a brochure or magazine cover which is in the form of camera-ready line copy. It should have only one plate carrying all elements, and any lettering should be done with transfer type or by hand. The design should contain at least two tints of noticeably different percentages, accomplished by means of shading films.

Shading sheets are used like cut-out transfer type. Peel the plastic film away from the protective wax-paper backing sheet and apply it directly on top of the previously inked art work. Be sure that the ink is dry and that there is no eraser dust or other debris on the surface. Next, cut out the section you desire to use with a frisket knife or sharp needle held in a pin vise. Then peel off the excess sheet, leaving the cut-out segment over the work. The film is then treated exactly like cut-out transfer type, burnished to ensure adhesion. Once it is rubbed down, it is invisible.

Avoid the temptation to overlap tints. To do so creates what are called "moiré" patterns (see Fig. 318), resulting from optical conflict between the two sets of dots. These patterns should be avoided in half-tone reproduction except in special circumstances where the kind of dissonance associated with Op Art or Acid Rock music is desired.[2] A standard solution to this exercise will be sharp, clean, and clear (Fig. 319).

FIGURE 317 Typographic designs employing transfer type

FIGURE 318 Moiré patterns result from the optical conflict between regular patterns of dots or lines superimposed upon one another

FIGURE 319 Design using shading sheets. Program cover for an international conference of ornamental ironworkers meeting in Vienna. Arressico, 1980.

FIGURE 320 Drawing done with rub-off transfer patterns. "RI": Fashion Figure

Rub-off sheets of benday dots and similar patterns are difficult to use for tinting of areas because the dots tend to "crawl" about and get out of line as you rub over them. They are, however, quite good for producing line copy that has a textured, spontaneous look.

There are also rub-off versions of shading sheets, but it is nearly impossible to keep the tiny dots in line when using them. It is possible, however, to achieve some very unusual and effective drawing effects with them (see Fig. 320).

EXERCISE 9–3: Doing a Mechanical for a "Square Halftone"

Materials: Illustration board
Frost-surfaced tracing film or heavy tracing paper
India ink
Glossy black and white photograph

Tools: Blue pencil
T-square and triangles
Red fiber-tip pen or red pencil
Rubber cement
Mechanical drawing pen (ruling or technical fountain pen)
Ruler

Any rectangular halftone is referred to by platemakers as a *square halftone*. The term indicates that the corners are at right angles, that it has been "squared up." (There are also round, oval, and silhouette halftones.) Select a photograph at least 8″ × 10″, and crop a detail 4 inches on the longer side to be used as a square halftone. Mount the photograph on illustration board and indicate crop marks in red just as you did for Exercise 4–3 in Chapter 4. On the margin write "square halftone." Draw *registration marks* in the top and bottom margins. These are cross hairs, frequently centered in circles (Fig. 321), used to ensure the proper positioning of art. The designer may draw them or may use preprinted plastic tape. At least two such marks must be applied to each sheet of the mechanical so that they are directly and *precisely* on top of those on the bottom sheet.

Place a sheet of tracing film over the mounted photograph, which is taped down on your drawing

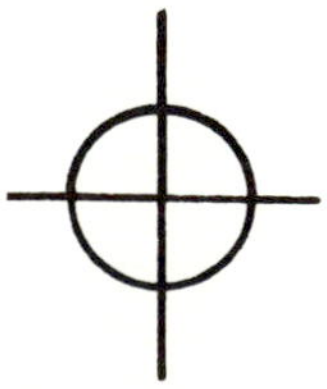

FIGURE 321 Registration mark

SQ HALFTONE

BUGGIN' OUT

75%

a

BUGGIN' OUT

b

BUGGIN' OUT

FIGURE 322 A: A mechanical
made for Exercise 9–3; B: As it
would look if printed

board. Fix it in place with masking tape and position registration marks directly over those on the illustration board. Using your T-square and triangle, draw light blue pencil lines connecting the crop marks. (Pale blue does not photograph and therefore need not be erased. On the other hand red photographs blacker than black and is therefore used for special instructions to the platemaker.) Following the box made by these lines, draw a rectangle of solid black as the ''window'' for the halftone. Using transfer type, give the photographic detail a title. Indicate the external margin of the design on the overlay by using the same kinds of strokes you do for crop marks; label these ''trim edge.''

Fig. 322 is an example of a completed exercise. Note it has a size indication in terms of a percentage,

in this case, 75 percent. In printing, the size of the original art is always taken as 100 percent and instructions to the printer are given in terms of that figure. A 75 percent reduction means that the art is to be reduced by one quarter. A 50 percent reduction is one half the size of the original ($8'' \times 10''$ becomes $4'' \times 5''$) and a 25 percent reduction is one quarter ($2'' \times 2\ 1/2''$). The percentages can be determined by use of a *proportional scale,* a low-cost device consisting of two discs of thin plastic calibrated to give precise ratios by simple alignment of two figures (Fig. 323). If you wish to know what percentage enlargement would result from increasing an $8'' \times 10''$ rectangle to 13 inches on the short side, the proportional scale will let you know that it is 162 percent and that the long side will measure 16 1/4 inches.

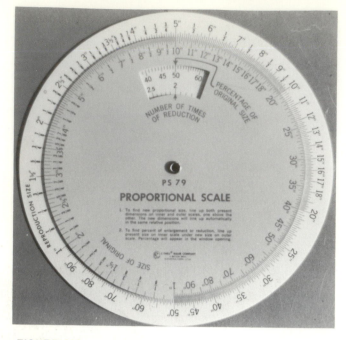

FIGURE 323 A proportional scale

EXERCISE 9–4: Combination Halftone, Dropout, and Surprint

Materials: *Illustration board*
Frost-surfaced tracing film or heavy tracing paper
Glossy black and white photograph, 8" × 10"
Transfer type (optional)
India ink

Tools: *T-square and triangles*
Pencils, pens, etc.
Burnisher if transfer type is to be used
Ruler

Prepare a mechanical using two overlays and a photograph mounted on illustration board to produce a printed result which will contain a square halftone combined with both dropout and surprinted elements. The halftone reproduction should measure 3" × 4 1/2", which means that you will also have to scale and crop the photograph. In Fig. 324 you see a photograph, the elements of the mechanical, and what would result from combining the elements into a plate.

FIGURE 324 A mechanical made for Exercise 9–4
A: The photograph taken in front of the Paris Opera
B: The line copy for the dropout and surprint
C: Printed version of the combination plate

a

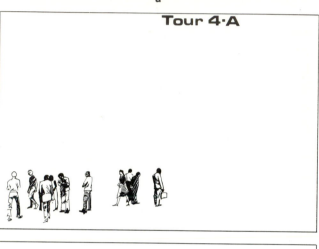

Tour 4·A

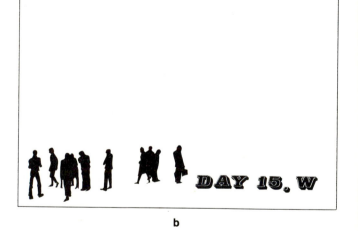

DAY 15, W

b

c

Tour 4·A

DAY 15, W

EXERCISE 9–5: Typeface Montage

Materials: *Large format fashion or homemaking magazines containing lots of ads and a large variety of typefaces*
White drawing paper
Rubber cement

Tools: *Single-edge razor blade or frisket knife*
Cutting surface

Make a design of one-inch squares no less than 6 inches on the short side and 7 inches on the long side. Each square should contain part of a letter or letters from the advertising in the magazine but should segment the lettering in such a way as to render it illegible. You may use as much or as little of a letter as you wish so long as it cannot be identified.

Work in monochrome only (that is, in variations of a single color). Two of the sample solutions in Fig. 325 were originally in black and white. The one containing the heart shape was red against light pink and the other was originally green against greenish black.

Choose each square for its individual merit and then arrange the squares in a pleasing design. You may find it necessary to return to the magazines for additional squares, of course.

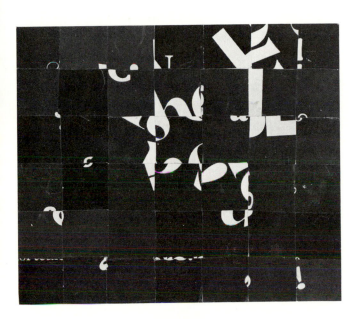

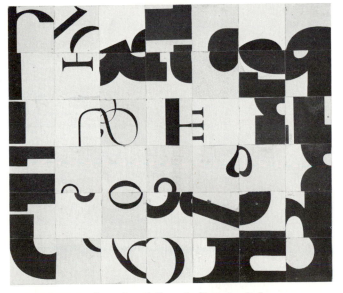

FIGURE 325 Sample solutions to Exercise 9–5

EXERCISE 9-6: Pen Lettering: Round Nib

Materials: *Graph paper with 1/4- or 3/8-inch*
squares
India ink
Tools: *Speedball pen B-5 or B-4 and penholder*

Pen lettering is initially difficult for most people to do and requires a good deal of practice to do well, but it can be mastered with patience and application. Speedball's standard manual for pen and brush lettering gives succinct, clear, useful instruction. These inexpensive little books are available at art supply stores.

Take a sheet of graph paper and copy the strokes of the sans serif alphabet (Fig. 299) using the following procedure: Ink the pen either by dipping it into the ink, removing the excess against the bottle rim, or by filling it. Make sure the paper is squared up with the table rather than slanted (unless you are doing italics). Then practice the strokes illustrated in Fig.

FIGURE 326 Strokes and directions for hand lettering

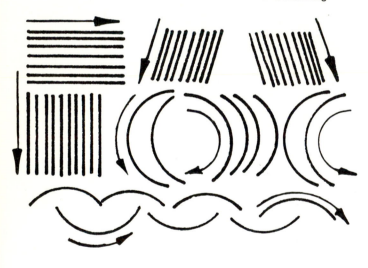

FIGURE 327 Sans-serif hand lettering on graph paper

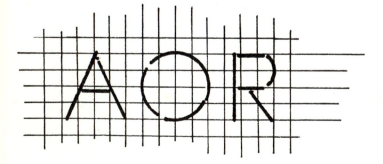

326. Always make strokes from left to right and from the top of the character toward you. (Curves require you to fudge this instruction slightly, but in general movements will follow this prescription.) When you have mastered the individual strokes, you can proceed to combine them into characters (Fig. 327).

Eventually, you will not need to use graph paper as a guide to proportions but will be able to follow lightly drawn horizontal pencil lines and letter forms sketched in lightly with a pencil. Since lettering pens come in a variety of sizes, the system is easily adapted to smaller or larger lettering, up to a point. After a certain size is attained, good sense dictates the use of brushes for poster lettering and for sign painting. The principles are the same as for pen lettering but facility with a brush is usually harder to attain than with pen and ink.

EXERCISE 9-7: Pen Lettering: Chisel-Point Nib

Materials: *Same as for Exercise 9-6*
Tools: *Speedball pen C-2 or C-3 and penholder*

Figure 328A discloses the variety it is possible to attain with the technique described in Exercise 9-6 but using a chisel-pointed pen, which yields strokes that are contrastingly thick and thin. The effect varies depending upon the angle at which one holds the pen. Held at a 45° angle it will produce what is the essential Roman letter style (Fig. 328B). (Incidentally, this letter style is never called sans serif in spite of the fact that it has no serifs.)[3] Careful addition of serifs with the pen held at an angle of 0° or 90°, depending on the position of the serif, will give you a hand-lettered version of the more conventional Roman (Fig. 328C).

The most comfortable angle for most people to hold a pen is somewhere between 15° and 30°, and chisel-point lettering based on pen lines within these angles has a natural flow (Fig. 328D). The extremes of horizontal and vertical are much harder to work with and not every letter form can be used without revision, but some decorative styles do grow from such exaggeration (Fig. 328E). These examples resemble typefaces called "Glamour Bold" and "Playbill" (Fig. 328F). The latter, however, can be drawn more effectively as built-up lettering in which multiple strokes constitute the character (Fig. 328G).

Experiment with various pen angles and alphabets, including *cursive* versions, in which the letterforms are connected as in script handwriting, and *italics,* in which the standard verticals are made to incline.

ABCDEHIJKMN OR STU W X Y Z

FIGURE 328

A: Chisel point work with the pen held at the following angles: 0°, 45°, and 90°

B: Roman style single stroke lettering

C: Roman style single stroke lettering with serifs added

D: PAUL GASTON. Congratulatory note. 1979. Felt pens on parchment, 9 × 11". Private collection

E: Lettering done with pen held at 0° and 90°

F: Glamour and Playbill typefaces

G: Playbill lettering done as built-up letters

b
ROMAN

c

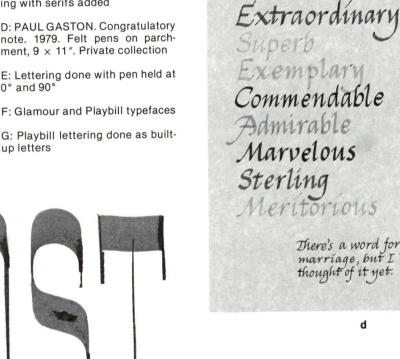

Remarkable
Splendid
Extraordinary
Superb
Exemplary
Commendable
Admirable
Marvelous
Sterling
Meritorious

There's a word for your marriage, but I haven't thought of it yet.

d

HOST
bar

e

Glamour
Playbill

f

g

EXERCISE 9-8: Built-up Lettering

Materials: *Same as for Exercise 9–6*
Tools: *Speedball pens B–2, B–5, and C–2, and penholder*

The characters in Fig. 298 are examples of built-up lettering based on the width of a B-5 Speedball pen. One stroke is the magnitude of a serif, two strokes the weight of the lighter elements, and five strokes that of the heaviest (see Fig. 329). Render the Roman alphabet using the same technique and maintaining the left-to-right, top-to-bottom stroking procedure.

Render the sans serif alphabet (Fig. 299) with a B-2 pen and square off the ends of the letters with a C-2 pen.

All of this could be done using drafting equipment to guide the tools, but such care is not really necessary for our purposes.

FIGURE 329 Roman alphabet done as built-up letters

EXERCISE 9-9: Mechanical Lettering

Materials: *White drawing paper*
India ink
Tools: *Ruling pen or Speedball B–5*
Drawing pencil (H)
T-square
Triangle
French curves
Compass

The block lettering in Fig. 300 is a very simple version of mechanical lettering. Fig. 330 indicates the procedures used in rendering the characters. Normally, one would use drafting pens or technical fountain pens to outline the forms in the first step of inking, but it is quite possible to do that inking with ordinary lettering pens if sufficient care is taken to keep the straight edges and curves a tiny distance above the surface of the art work. This can be done by taping small pieces of cardboard to the underside of the T-square and to both sides of the triangles and curves (since you will want to flip these over). A scrap of

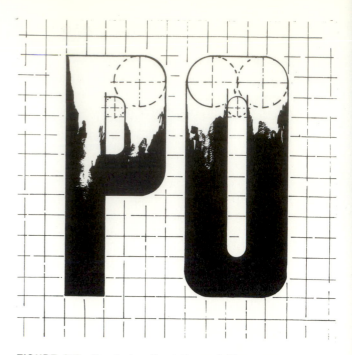

FIGURE 330 Rendering the letters of Figure 300 mechanically

matboard can be cut up and turned to this use very swiftly and at no expense.

Obviously, care must always be taken that lines are not smeared when wet. Also, it is wise to keep your guidelines very pale so that they can be easily erased. Never dip your pen into the ink when doing this kind of work; drafting pens (technically ''ruling pens'') are always filled from the bulb stopper of the inkwell anyhow, but do the same if you are using lettering pens.

All of the lettering in Figs. 298, 299, and 300 can be done mechanically. Establishing the curves is correspondingly more difficult in the cases of Figs. 298 and 299 than in that of 300 but the overall approach is the same. You might try doing a few of the characters in the Roman and sans serif styles.

EXERCISE 9-10: Grid Layouts

Materials: *White drawing paper*
Picture magazines such as Life, *Smithsonian,* Playboy, People
Rubber cement
Tools: *Single-edge razor blade or frisket knife*
Ruler with a metal edge
Cutting surface of chipboard, hardboard, heavy cardboard or similar material

Do five layouts, using the same grid for all of them. Do not use the six-unit pattern like that chosen for Figs. 312 and 313. Each design should employ a different approach:

1. Alleys crossed, but not gutters or margins.
2. Illustrations bled.
3. Some illustrations butting the gutter or gutters.
4. Illustrations butting gutters and also bleeding (but not all necessarily doing both).
5. Irregular, silhouetted forms.

The layouts should be for a two-page spread, each 8 1/2″ × 11″.

Cut illustrations and blocks of type from the magazines; select pictures that will not look like silhouettes when pasted on white paper. Choose black type of an appropriate size and character. Use black and white reproductions for all but one of the designs. You may crop the illustrations (that is, cut into them to achieve the right shapes and sizes) and the blocks of type need not constitute whole paragraphs or say anything pertinent to the illustrations.

Don't concern yourself with meaning; the pictures can be unrelated and can have nothing to do with the captions. Once you have determined an arrangement and cut out suitable pictures and blocks of type, paste the pieces onto the paper with rubber cement.

If you are careful and relatively neat, you will end up with designs that are far more professional-looking than you might expect. The geometry itself produces a pleasing set of relationships and the layouts have an equilibrium that is appealing. The continuity of elements from page to page is strong. Inevitably, you will find that your result looks not only pretty good, but also familiar. That is because so many graphic designers use grids to generate shapes. As Bud Donahue says in his book, *The Language of Layout:* "Acceptable results come quickly and easily. So what is the big objection? They could have been done by a computer."[4]

notes for chapter 9

[1]The word *folio* has other meanings in the graphic arts. In addition to being the slangy shortening of "portfolio," the term refers to a large sheet of paper folded in half so that it forms two leaves (four pages) in a folder or a book. The word also denotes a book of the largest size, normally one larger than 11 inches or 30cm. tall.

[2]Moiré patterns commonly occur when halftone reproductions are made from copy that is already in a halftone or tinted form. This is an especially common problem in *gravure* printing, which requires that even line copy be turned into halftone. Slight errors of registration and unfortunate screen angles can produce moiré effects in multicolor halftone reproductions, too.

[3]This is one of the many inconsistencies in the classification of lettering styles. Another is that some typographers use the term "Gothic" as a general description of most sans serif letter styles, while others reserve the term for what is frequently called "Black Letter" (known popularly as "Old English") or "Text."

[4]Bud Donahue, *The Language of Layout* (Englewood Cliffs, N.J.: Prentice-Hall, Inc., 1978), p. 37. Mr. Donahue, whose book is an extremely valuable guide to the craft of layout drawing, is an experienced advertising artist with a rather low "threshold of indignation." He even has a chapter entitled "Let's Banish the Grid." Clearly, the authors do not share this objective. But Donahue is refreshingly knowledgeable and is as candid as he is acerbic.

designs for dwelling spaces and abiding places

Virtually anyone reading this book will be able to look up from it and find something in sight that is neither living nor produced by industry. A painting perhaps. Or a ceramic, a drawing, a piece of hand-woven fabric—maybe even a piece of handmade furniture. Despite the presence of such things within one's living space, though, the predominate features of the environment are almost surely mass-produced. Even the doors to the room and the lumber for the jambs were made by machines. And the wooden things are the most natural objects in sight; that is, their substance has not been much altered, only their form has been changed. Everywhere else manufactured materials such as steel, aluminum, and synthetic plastics have been cast into appliances and furniture. From the beginning of the Industrial Revolution hand-woven fabrics have been relatively absent from ordinary homes. Even those of us who sew today make very few items for ourselves to wear. Today the designer who does handwork with fabrics is likely to be neither a tailor nor a seamstress but an artist whose fiber creations (Fig. 331) have more in common with fine art than with practical craft. Virginia Jacobs's *Cakewalk* is a poncho-like garment, but its particular interest for us is in the way the beads, quilting, and applique elements have been brought together in a rich overlay of intertwining elements reminiscent of folk art. Charlotte Patera, in her quilt (Fig. 332), is more explicitly derivative of folk art forms; hers is a contemporary North American version of the Molas made by the Indians of Central America. Both the Patera and the Jacobs look as if they have some family connection with Ron Thomas's obsessively detailed, fanciful painting (Fig. 333).

Our fondness for handmade, irregular things is partly an expression of humanity's vast but nonetheless limited tolerance for the monotony of geometrically rigorous shapes and synthetically homogenized surfaces. We even manage to program irregular variations into what is otherwise uniform by, for example, producing plastic laminates that contain illusionistic wood grain. Too, the competitive merchandising of products in the United States, Western Europe, and Japan provides a certain amount of variety among the items available for sale.[1]

Americans' desire to escape the depersonalization of mass-produced building materials was among the impulses contributing to put, as one colleague has said, "the gingerbread on Grandma's front porch."[2] Houses like those in Fig. 334 possess a certain picturesque appeal for many people today for precisely the reasons they were liked by our Victorian forebears; the clutter of columns and arches and pinnacles set one upon the other seems far removed from the dreary practicality of the standard dwelling place. Moreover, as many who have renovated the old houses know, they have an unwarranted reputation for impracticality. The protruding bays of the home in this photograph give virtually every room an excellent exposure throughout the year. *Solar gain* during the winter softens the interior chill. Heavy draperies and blinds drawn down in summer team up with the vented attic to produce a natural flow of cooler air from the ground floor upward, providing natural air-conditioning on all but the very hottest, most humid days. Of course, space was cheap in the days when these homes were built to house the prosperous burghers of the city and that's the main appeal of the buildings today. Insulation and double

FIGURE 331 VIRGINIA JACOBS. *Cakewalk.* © 1980. Applique, quilting, beading, 56 x 102″. Private collection

There are art works using traditional sewing techniques as a medium of expression. In today's world, machinery and assembly-line techniques take care of most of our practical needs, such as the provision of clothing and bedcovers. The handcrafts of sewing and tailoring are likely to be practiced by fine artists or specialists working in the world of fashion for an elite clientele.

FIGURE 332 CHARLOTTE PATERA. *Molistic.* 1980. Four Indian Mola techniques (applique and reverse applique), 37 x 39″. Private collection

FIGURE 333 RON THOMAS. *Rainbow Man.* 1975. Acrylic on canvas, 50 × 49½″. Collection Southern Illinois University at Edwardsville.

FIGURE 334   House from the Victorian period

Victorian houses are frequently overloaded with decorative lathework and jigsaw ornamentation but they just as often have floor plans that are very effective in catching the sun and giving the inhabitants an eye on the world through bay windows. Frequently, they had natural drafting to keep them tolerable in summer and were designed to reduce living space needing heat in winter by allowing whole sections to be closed off during parts of the day or season.

glazing can make a well-built nineteenth-century house quite practical today, even with elevated costs for heating and the new convenience of air-conditioning.

The so-called "energy crisis" was not a concern for most of us when Ravine House (Fig. 335) was designed and built, although this home is sited to take advantage of a low-flying winter sun through the great windows of the deck and is built so that the 6-foot overhang of the roof sheltering that deck screens out the rays of the summer sun. While it is much larger than it appears in photographs, this house is more compact than any comparably middle-class Victorian domicile (see Fig. 336). Since it is a unique design rather than a standardized pattern, it sacrifices some square footage in favor of style (Fig. 337). It is a *kind* of style many do not care for—a sort of "rustic Modern" that is as much a period piece in its way as the Victorian houses are in theirs.

The term *Post-Modern* first appeared in connection with architecture, and it has usually been applied to buildings that are modernistic in the way certain commercial architecture is. An interesting and also descriptive contrast between Modern and Post-Modern occurs in the work of architect Philip Johnson (b. 1906). During the late 1950s Johnson collaborated with Ludwig Mies van der Rohe on what has

come to be thought of as a paragon of Modernist skyscraper design. The Seagram Building (Fig. 338) in New York City is a bronzed gridwork of steel and amber glass rising straight up from an open plaza on Park Avenue. This design is a classic of the Bauhaus approach to architectural design, and the relationship of part to part has been carried through with amazing and unsparing rigor. Johnson's later design for the building housing the main offices of AT&T (Fig. 339) is a startling departure from the tradition with which he had been long connected. It does not look old-fashioned, really, because the sheer walls and window proportions are clearly contemporary in character. Expectedly, however, the broken pediment silhouette and great archway portal have caused this building to be dubbed "Chippendale Modern."

In domestic architecture, the Post-Modern style has been practiced most notably in the United States by a man who first gained attention as the premier Post-Modern critic of architecture, Robert Venturi (b. 1925). Some notion of his approach to the problem of sheltering human beings can be divined from a glance at his beach house in Loveladies, New Jersey (Figs. 340 and 341). It looks like nothing more than a two-tone box with queerly spaced windows, but it resembles some historically important buildings of the

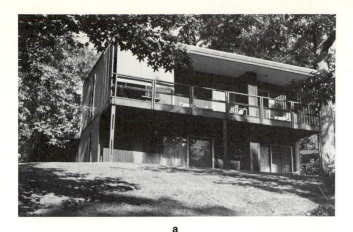

a

b

FIGURE 335 JOHN ADKINS RICHARDSON. Ravine House. 1966. Edwardsville, Illinois A: Southwest exposure, summer; B: Southwest exposure, winter; C: Entry deck

Ravine House is an example of small-scale domestic architecture in a Modern style. The facade of the building overlooking the heavily wooded vale is mostly double-glazed window surface opening in summer (A) from a shaded deck and patio, and serving as a primitive passive solar collector during the winter (B). The house is angled into the hillside so that the lower floor —containing a studio, two studies, and a garage—is, in effect, an earth-insulated, walk-out basement. The main entrance is from an entry deck (C) on the upper level.

c

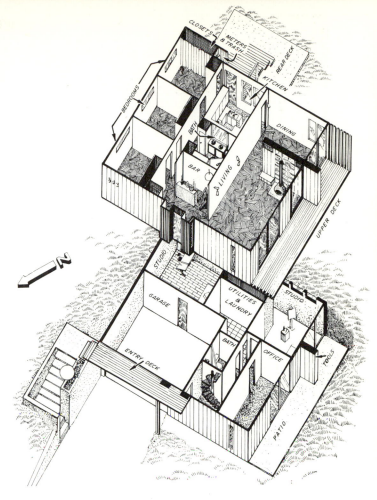

FIGURE 336 Sectional diagram of Ravine House.

This rendering of Ravine House shows the way in which the utility core serves as a central focus of the design. Compact arrangements of this kind are both economical and efficient.

FIGURE 337 Living area of Ravine House.

Nonstandard items such as the firepot and the divider wall which separates living from dining space and extends out onto the deck for storage are far costlier than any practical advantage warrants. Purely stylistic uniqueness of this kind nearly always costs extra in human effort and money. In this case, the cost was subtracted from the space of the bedrooms, a sacrifice the owners didn't mind but one that many people might not wish to make. ("Flat" roofs also run costs up; pitch roofs are cheaper because no roof is truly flat and gauging slight inclines and building up the roofing to resist moisture is more difficult than using standard rafters and an angle that sheds water.)

FIGURE 338 LUDWIG MIES VAN DER ROHE and PHILIP JOHNSON. Seagram Building, New York. Designed 1958

FIGURE 339 PHILIP JOHNSON and JOHN BURGEE. Design for American Telephone and Telegraph Building, New York. 1977

FIGURE 340 ROBERT VENTURI, JOHN RAUCH, and SCOTT BROWN. Lieb House, Loveladies, N.J. 1967. Photo credit: Stephen Hill.

FIGURE 341 Interior of Lieb House

early Modernist movement, such as Le Corbusier's Villa Savoye (Fig. 342), in that the exterior provides a simple shell for something more complex inside.

All of these houses—the Victorian manses, the Ravine House, the Villa Savoye, and the Lieb House—are highly distinctive, individualized dwellings specifically designed for people with definite preferences. In terms of the great sweep of history they are trifling self-indulgences of people willing to pay a certain price for being slightly different. Most people in America have been willing to be housed in mass housing in the form of single-family dwellings like the ones in Fig. 343. Their houses may have been larger or smaller than these, more modernistic-looking or less, the cost higher or lower, but the domiciles were essentially the *kinds* of houses we see in the photographs. Despite all that has been said against these ordinary houses by succeeding generations of designers and architects, they've done pretty well by their owners. Usually, their value has increased over the years, providing a base for families to progress up the scale to more expensive dwelling places. They have been fairly comfortable in most seasons and have provided sufficient space for most of the owners, particularly in contrast with the common places of Europe and Asia. Now, however, it seems as if their long-predicted demise is close at hand. Already, the sheer costs of building such homes are shrinking them in size and at the same time driving average buyers to extreme measures in search of

FIGURE 342 LE CORBUSIER. Villa Savoye, Poissy, France. 1928–30

The Seagram Building is an example of Nordic Modern architecture in the so-called *International Style.* The AT&T Building, done by one of the same architects, is an example of what has come to be called "Post-Modern" architecture. This unfortunate term is so anomalistic as to be meaningless but is obviously intended to suggest a contemporary mode that has some things in common with the Modernism preceding it while being less serious and less concerned with bare-boned functionalism and engineering logic.

The design for a beach house in New Jersey is by an architect and critic whose name is virtually taken as a synonym for Post-Modernism. Robert Venturi uses the wall and windows as a sort of ornamental mask for the interior functions of the house. This is definitely not what the "form follows function" tradition of Modernism requires. Yet, the Venturi beach house is not altogether unlike a major classic of early Modernism, the Villa Savoye by the French master Le Corbusier (Swiss-born Charles-Edouard Jeanneret, 1887–1965). The upper level of the Villa Savoye is a box containing a beautifully articulated interior of several levels tied together and attached to terraces by ramps. A good deal of the upper story is open to the sky, including the roof garden shielded from the wind by one of the cylindrical towers. Thus, while the Corbusier design may be somewhat more functional as well as more elaborate, it also conceals rather than reveals its nature with a skin of windows and walls. The similarities and differences of the two houses—like those of the Seagram Building and AT&T headquarters—show why the Post-Modern styles are not the same as "non-Modern" or "anti-Modern" might be.

FIGURE 343 Some American streets

Here are four fairly typical American neighborhoods, each built up at a different period but all photographed during the same year. Time has produced many changes in the older ones. Changes yet to come among the more recently built will make them too less alike to the casual passerby. However, the houses on a given street will remain remarkably alike in floor plan and substructure. An occasional eccentric may pop up, just as a particular family may not fit in with others on the block. Still, the tendency has always been toward enormous conformity of people living in an area, in terms of income, education, position, and home style.

financing. Inevitably, a renaissance of clustered homes—apartment houses and condominiums—is in emergence. The innovations typical of these buildings, however, are on a par with Arressico's "Arressicondo" design (Fig. 344), an attempt to nest split-level townhouse apartments together in such a way that the utilities are compressed into a minimum space while the living space is glamorous in effect and the exterior space offers a choice of decks, roofed patios, open back yards, or carefree airspace. This makes for greater efficiency and a more individually styled set of apartment units, but it does very little in the way of providing genuine options for society at large.

FIGURE 344   ARRESSICO. Arressicondo design. 1982

A condominium design attempting to provide multiple, split-level dwelling spaces with various kinds of "back yards." In some units the living and dining space is on the second floor, in others on the entry level.

Uniformity of style in new developments and among condominiums appalls designers, but as decades pass, the neighborhoods begin to take on the appearance of diversity, albeit a diversity of detailed differences within a general sameness of quality and scale. This is not a serious matter to most people. Generally, the citizens of the United States are comfortable with conformity so long as it is not absolute or overtly enforced by civil authority. Tension between the desire to conform as part of a group and at the same time to be an individually distinct presence within that group is a constant in America. Every generation of adolescents seems to be uniformly coiffed and garbed to adult observers, but teenagers themselves know that their garments and hair styles are at least as unique as their parents' clothes or homes.

We go into this matter of perceptions of similarities and differences because we are about to embark upon a discussion of modular building units. Throughout this book modules in one form or another have received considerable stress. Modules are very useful, and nowhere more than in mass housing. Unfortunately, groups of them are apt to seem

hivelike when a great number of units are brought together in a complex, as they were for Moshe Safdie's Habitat (Fig. 345). This pile-up of one- and two-story apartments consists of precast concrete modular boxes which were fitted together by an assembly-line technique. It looks strange and impersonal, but its inhabitants like it, for Habitat is more like a Mediterranean seacoast village than an apartment house. The cubicles have been designed so that they can be coupled together in a number of ways, then opened up and partitioned to make individual interiors more distinctive than are customary in ordinary apartments or single-family dwellings. Moreover, each apartment is sealed off from the others, having its own entrance from "sidewalks" that form a network of streets through the cluster, and each apartment has its own private terrace. To outsiders the place looks weird and confusing, the living spaces depressingly alike while horribly jumbled. To those who dwell within, however, it is evident that—as in the case of teenagers' clothing—apparent uniformity conceals great differences. In any case, the Habitat apartment units are no more alike than are the homes in any of the pictures in Fig. 343.

FIGURE 345 SAFDIE, DAVID, BAROTT, and BOULVA. Habitat, EXPO 67, Montreal, 1967

This complex of building units was constructed for EXPO 67 in Montreal as an experiment in low-cost, high-density housing. It turned out to be very expensive because of cost overruns, but it could be duplicated today for about a third of what it cost to build.

It seems evident that a main task for the designers of coming generations will be to invent dwelling spaces that are far more efficient in terms of space consumed and energy expended than conventional houses have been. Virtually everyone who has looked at the matter seriously agrees that the world faces two fearsome crises in the foreseeable future: recurring energy shortages and population growth. The first is merely a subsystem of the second, since more people usually consume more energy. Presently, the citizens of the technologically developed nations consume vastly greater amounts of energy per capita than the enormously increasing populations of the so-called Third World. In a very real sense, our progress has come at the expense of the peoples of underdeveloped Africa and Asia since, as ecologists never tire of pointing out, conventional sources of energy are finite. When we take more than our share of the planet's fixed resources merely because of our historic good fortune (having been born where the raw materials are and into a group whose ancestors managed to prevail over other peoples and natural obstacles), we are, in effect, consigning the greater number of human beings in the world to famine, drudgery, and death. If you could live in comfort on a third of the energy it now takes to run your home, the remaining 66.7 percent would be available to two other families on what Buckminster Fuller (b. 1895) calls our "Spaceship Earth."

Mr. Fuller once ran his "World Game" from the university where the authors teach, and on that campus is a building featuring one of his famous domes (Fig. 346). Fuller is a universalist who has devoted considerable attention to the problem of human survival on the planet Earth, which he conceives of as an enclosed ecosystem flying through space, powered by the sun it orbits. He feels that all architecture of the past is obsolete. He was trained not as an architect but as an engineer, and his ideal structures from the past are not buildings, but plant stalks, crystals, tension bridges, sailing ships, and airplanes. Strength, efficiency, and lightness dominate his designs, the most famous of which is the *geodesic dome*. A geodesic is the shortest distance between two points on a mathematically derived surface such as a plane (where the geodesic is a straight line) or sphere (where the geodesic is the arc of a great circle, that is, a circle that is like the equator on the

FIGURE 346 BUCKMINSTER FULLER. Religious Center, Southern Illinois University at Edwardsville, 1971.

Earth, where the radius of the sphere and circle are the same). One of the largest of Fuller's domes was the one used in the United States Pavilion at Montreal's EXPO 67 (Fig. 347). This dome was constructed of steel pipe and transparent acrylic panels; it measured 200 feet from floor to the highest point and 250 feet across. Altogether the metal pipes weighed 720 tons. Yet, in terms of total space contained by the structure, this averages out to four ounces per cubic foot! That is an absolutely astonishing figure. No other shelter constructed by human beings comes close to the efficiency of the geodesic domes. Built of triangular sections, they can be constructed of anything from cardboard to steel and they are tremendously strong, since they are derived from "unfolded" tetrahedrons—pyramids composed of equilateral triangles—the strongest known geometric form. There seems no limit to the size these domes might attain, though the largest to be built thus far is the Union Tank Car Company Dome in Baton Rouge, Louisiana. It is nearly 400 feet in diameter. Fuller has proposed a two-mile-wide dome to enclose New York City and has designed a 200-story floating, tetrahedron-shaped city for a site near Tokyo. This tetrahedron is composed of smaller ones and Fuller envisions each unit and any stable collection of the units as portable.

So far as energy conservation is concerned, Buckminster Fuller has many ideas on how to generate electricity from wind and sun and the surging motion of the sea. It should be understood that his architecture already addresses the problem merely be-

FIGURE 347
BUCKMINSTER FULLER. United States Pavilion, EXPO 67, Montreal, 1967

Two examples of Fuller domes. The first has been ornamented with the oceans of the Earth so that, from outside, it seems to be a normal sort of map and affords worshippers an "inside view" of the globe from within the sanctuary. The United States Pavilion was constructed on a far larger scale for the world exposition in Montreal in 1967.

HEX

5'2⁵⁄₈"

4'6⁵⁄₁₆"

PENT

4'6⁵⁄₁₆"

4'6⁵⁄₁₆"

5'2⁵⁄₈"

HALF-HEX BASE UNIT

a

b

FIGURE 348
A: BUCKMINSTER FULLER. Domes design for *Popular Science,* 1972. Courtesy of Buckminster Fuller.
B: A combination of geodesic domes

For *Popular Science* magazine's 100th anniversary, Fuller designed a do-it-yourself dome that we have redrawn as Fig. 348A, showing the three "hex-pent" units from which it is built, the way the web of struts would appear from within, and the exterior appearance. The exterior drawing emphasizes the way the modules would fit together by showing their seams in a heavier line, and the young woman and her dog are included for scale. Additional domes could be joined together to create the kind of home shown in Figure 348B. (Plans for the dome are available from *Popular Science* magazine.)

cause it accomplishes so much with such economy of means. For, when one considers energy consumption by a society, one must think of how much gas and electrical power is expended to, for example, mill a single finished two-by-four from a log that was once a fir tree in a forest. Energy expenditure cannot be restricted in meaning to what it takes to heat or cool a house but must also be calculated in terms of what it took to build the house. Human physical effort can sometimes become a part of that estimate, of course.

Like sea and wind and solar power, it is a generally renewable resource.

It is entirely possible to build your own "Fuller dome" from scratch, using relatively inexpensive materials and a plan devised by Fuller himself for the do-it-yourself readers of the 100th anniversary issue of *Popular Science* magazine. This vernacular dome is constructed from triangular panels of fir plywood held together by struts of standard two-by-four stock (Fig. 348A). To build the dome correctly, you will need more information than our diagram provides because it does not give details of doors, windows, foundation, and so forth. No matter what materials you might use, Fuller space is very inexpensive per cubic foot when compared with a conventional building made from the same material on a similar scale. True, the space is rather odd space to most of us and it would take a lot of adjusting to. Too, local building codes and covenants may make it impossible to put up such a structure on some properties. Also, the slant of the cuts must be precise and practically nothing in the design is sawed to a 90° angle. You would need accurate power tools and a determination to be careful, but any conscientious man or woman should be able to construct the dome in a couple of weeks. One of the most appealing things about it is that the parts can all be precut in a woodshop and transported to the site for assembly.

Obviously, it would be possible to join a number of the domes together into a larger interconnecting modular complex, the triangular modules forming pent modules and hex modules which, in turn, make domes that become modules from which a cluster might grow like a crystal (Fig. 348B). Among the difficulties dome builders have experienced in constructing assemblages of multiple domes is that they are hard to put together, even singly, so that they don't leak. Geodesic domes must always be caulked and no one ever seems able to do every seam and all of the gaskets just right. This is not to say that the problem is insuperable—it's just tiresome—but merely to note that even geodesic domiciles have their problems.

For all of their positive qualities of strength and lightness, the domes raise the same kinds of problems for building contractors and manufacturers of prefabricated dwellings that they do for the home-handyperson: Building codes and the inspectors who enforce them are locked into 90° architectural notions. The neighbors may not be prepared to tolerate a bunch of big faceted balls on the lawn next door either. And assembly is never *quite* as simple as Fuller's disarming stories about illiterate Indians plugging color-coded rods and panels together would imply. Still, these are not completely overwhelming problems in the United States, where we have a continuing tradition of homestead tinkering and eccentric practicality such that even conservative taste cannot resist positive ingenuity.

The neighbors of David Boulds (b. 1948), whose solar home (Fig. 349) stands on an ordinary street in the Mississippi River town of Alton were overjoyed to see what their local firefighter had wrought in his spare time. No commonplace do-it-yourselfer, Boulds had been a carpenter and electrician previous to undertaking an Associate Degree in Fire Science, and all three specialties supported his decision to renovate a decaying old hulk on Brown Street. After ten years, the home has begun to take on the contemporary look Mr. Boulds and his wife, Alice, prefer. Practically the entire front facade of the house has been transformed into support for a large

FIGURE 349
DAVID BOULDS. 3112 Brown Street, Alton, Illinois, 1981

A do-it-yourself renovation project that has transformed a decaying old house into a contemporary solar home on Brown Street in Alton, Illinois

solar collector. Even this is David Boulds's invention. He fabricated the modular cells of the collector from corrugated tin roofing painted with a very flat black paint to increase their capacity to absorb light and heat. The greenhouse effect of the double-glazed skylight not only builds up temperature but also prolongs the life of the heat by capturing it. The only power required to heat this house when the sun is out on sub-zero days is a small blower used to accelerate the flow of warm air through the furnace ducting that forms the substructure of the collector. Even when the system is completely passive, with the blower off, the building maintains an internal temperature most northern Europeans would find quite comfortable, in the 65° to 70° range. Air enters the system on the upper right corner of the solar collector, flows downward and into the next cell, where it is drawn up and around into the third, and so on, passing through every inch of ducting until it reaches a level of 185°. As it flows into the house through the registers, the air still measures 140°, more than adequate for winter comfort. Of course, it is important that the heat be retained, not dissipated, so this house is, like all energy-efficient homes, heavily insulated. Boulds has used 4 inches of foam in his walls and, in the attic, 11 inches of cellulose/fiber glass batting. The doors are themselves foam-filled; in winter the outside of the door may be freezing to the touch, but the inside will be as warm as it would be in summer. A gas furnace backs up the solar system in the event of overcasts. So far, it has hardly run at all. The heavy insulation and tightly caulked seams of the structure have made it unnecessary to install air-conditioning; up to now, an attic fan has been sufficient to cool the entire place in the summer—and this in a region where temperatures and humidity in the high 90s are altogether common.

A system like Boulds's could be linked up with a similar one devised by famed ecologist Barry Commoner (b. 1917) for heating water. Instead of flat black compartments of tin, Commoner uses pans of dark wax. Rather than ducts, copper tubing is run back and forth through the wax. When oil is pumped through the course of tubing, the heat collected by the wax is transmitted by direct conduction through the copper to the fluid. When the oil is carried through the tubing as it encircles a tank of water, the water becomes steaming hot.

David Boulds also helped an acquaintance design and build a home that is mostly underground. Only the roof elements and entrances are visible from the driveway. When it comes to insulating, the *R-value* of earth is not hard to beat, but earth *is* dirt cheap.

One of the authors lived as a child in a "sod house" (Fig. 350) on the prairie of northern Wyoming. The place was made from 8-inch blocks of grassy soil stacked up into walls shored by wooden posts and beams, then roofed over, and the roof in turn layered with sod. Such a structure is much stronger and more resistant to erosion than you might imagine, because the roots of the prairie grasses are strong, intertwining tendrils that lock the soil within a wirelike grip. The house was as cool as a root cellar in summer and tolerably warm in a climate where winters are 40° below zero and the wind

FIGURE 350 A drawing, made from old photographs, of a sod house that once stood on the prairie between Upton and Moorcroft, Wyoming.

FIGURE 351 An underground house design. Arressico, 1981. Essentially, the house is a large concrete basement with solar-assisted radiant heating coils in the floor and heavily insulated roof area at the access level.

moves like a scythe across the plains, tearing at sagebrush and laying frozen grass out flat. The principal problem with such sod houses is that they are a bit *too* ecologically balanced. The grass growing on the walls and roof is good protection from wind and rain erosion and cactus flowers on the roof are picturesque, but roots reach in through walls and insect life goes on as well.

Also, for earth to really insulate a roof, it has to be at least 2 feet deep. That's enough to collapse a

FIGURE 352 TED BAKEWELL III and MICHAEL JANTZEN. Autonomous Dwelling Vehicle, 1979

The Autonomous Dwelling Vehicle resembles what a lot of us think a UFO should look like. The power for electricity is generated by the solar photovoltaic panels and reflector, which extend out from the vehicle's side decks. (The decks and panels fold up out of the way when the craft is traveling.) The skylight bubbles on top admit light and are vented to allow hot air to exit. A corrugated fiber-glass "tunnel" between the two skylights covers pipes that carry the heat-exchange fluid used to heat water.

wood-frame house, even without any snow. Still, these untoward eventualities can be overcome in the simplest possible way: build your house just as you would a basement, as a reinforced concrete box surrounded by soil (Fig. 351). The military have been doing this with fortifications for about a century and there is no reason to ignore a practice which could protect many American midwesterners and Soviet steppelanders from tornados and could reduce fire hazards significantly.

Many people, however, are discouraged by the idea of living a subterranean existence, even when assured that they can have some windows on a side or level of the house and that they'll save a good deal of money by living in a bunkerlike dwelling. They might consider the Autonomous Dwelling Vehicle (Fig. 352) by Ted Bakewell, III (b. 1946) and Michael Jantzen (b. 1948).

The Autonomous Dwelling Vehicle can go anywhere. It looks like an alien space probe come to Earth from a mother ship hiding on the dark side of the moon, perhaps, but the only way it can fly is by helicopter airlift. It's built to the same scale as existing luxury vans and can be moved on the highway by truck, on rail by freight car, and across seas by freighter. It is also adaptable to flotation as a vehicle on its own and could be moved on canals as on highways. Such versatile mobility makes the ADV useful to Bakewell, who actually uses it as a residence on the building sites of his family's construction firm, but the mobility is not what makes the home interesting. What brings the ADV to our attention is its lack of reliance on utility grids or fossil fuel.

The experimental ADV design is self-sufficient in energy, ecologically neutral, cozy to live in (Fig. 353), and economically competitive with vans of a comparable size.[3] Electricity for the fluorescent lights, air conditioner, and heating are supplied by a wind generator and by ARCO Solar *photovoltaic* panels augmented by a reflector. Each photovoltaic cell generates 20 volts of electricity directly, without any loss through mechanical friction, and stores it in a magazine of four marine batteries. This system is further augmented by components tapping other energy sources—by, for instance, two fan-assisted air collectors built into the exterior skin and by Thermol 81 heat storage cylinders under the bed compartment. Thermol 81 will store sufficient heat to keep the vehicle at 68° for 48 hours in zero weather or at 68° from Monday through Friday on days that are 30°. A greenhouse, controllable skylights, and an airlock entrance provide still more passive solar gain, and a swiveling fireplace/wood stove and small heat pump complete the air-warming system. A waste incinerator joins with a solar collector to provide hot water. There is on board a rain collector, a water purification system, and, for nonburnable wastes, a Clivus Multrum, the Swedish toilet which composts human wastes odorlessly. Bakewell can use a blender in his vehicle; a foot pump and the wind generator compress air so that a universal air motor can power the blender, as well as rotating shaft tools. Everything is used as conservatively as it might be in a space capsule. The owner showers in waste water that has been purified and disinfected. The showerhead is a nozzle developed for crop dusting; it produces an extremely fine but effective spray mist. Some things cannot, it is true, be accommodated in so self-sufficient a domicile; such energy-intensive appliances as electric ranges, dishwashers, and washing machines draw far too much power. Bakewell uses small alcohol-burning and solar stoves for cooking. Dishes are done by hand, naturally. The laundry could be done that way, too, but Bakewell usually takes it out.

What is perhaps most amazing of all about the prototype ADV is that it was not built by a team of foundation-funded technicians, but by Bakewell and Jantzen themselves, without any grants or subsidies. These young men have studied architecture, engineering, and design. Michael Jantzen is a professional multimedia sculptor and conceptual artist as well as a building designer. He ran an experimental energy laboratory at Washington University. Also, Michael always works in concert with his wife Ellen, who is a trained artist and designer, a seamstress, and a professional gardener whose work has been the subject of magazine articles. She does all of the sewing for the designs and grows for sale vegetables and

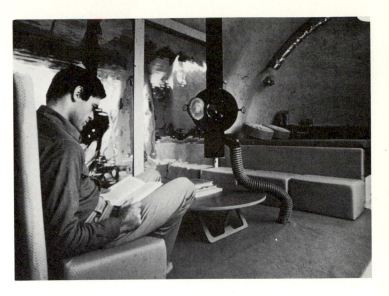

FIGURE 353 Interior of the Autonomous Dwelling Vehicle

The very striking sculptural form in the center of the ADV interior is a tiny fireplace-like wood stove made by Jantzen from a 13-inch Freon gas cannister. It has a screened opening 6 inches wide and can be swiveled to face any direction. Partly because the flexible hose draws air from beneath the structure instead of from within it, this stove is capable of heating the entire space in just a few minutes since none of the interior warmth is swept out through the chimney. Ellen Jantzen's decorative pillows are also insulating "plugs" which can be inserted in the skylights on cold winter nights.

herb seedlings raised in a Jantzen-designed solar greenhouse (Fig. 354). The garden is tracked for yields of varieties on a home computer. One might expect such a couple to own an unusual home, and they do (Figs. 355 and 356), even as they are engaged in constructing another. Ted Bakewell says: "Michael's house is the mother ship of the Autonomous Dwelling Vehicle."

FIGURE 354 MICHAEL JANTZEN. Solar Greenhouse. Carlyle, Illinois, 1980

FIGURE 355 MICHAEL JANTZEN. Jantzen House, Carlyle, Illinois, 1977

As of this writing, the Jantzens live in this ecology-minded home that uses a computer to tell them how it is functioning. The south side of the house, seen in this photograph, contains two solar heating systems, one passive and the other active. The passive system depends upon the bubble windows on center which permit sunlight to fall upon a brick floor that radiates heat into the room. The active system is visible externally as two broad stripes on either side of the window. These are identical to the rest of the corrugated steel exterior but are painted black and enclosed in fiber glass. The two "eyes" at the ends of the structure are blister windows lidded with steel visors.

FIGURE 356 Interior of the Jantzen House

The interior of the Jantzen house. Notice the Brunswick folding chairs hanging from the wall. Not only do these highly durable pieces of furniture free floor space when they are not in use—in the old American tradition of Shaker furniture—they have been used here as decorative elements in place of pictures.

Even as a college undergraduate, Michael Jantzen had a genius for turning vernacular material to previously unheard-of uses. His buildings are masterpieces of serendipity; they use existing material in surprising ways. Like parts of the ADV and the earlier Solar Vacation Home (Figs. 357 and 358), the essential exterior shell of the new Jantzen home (Fig. 359) is made of prepainted steel agricultural silo components, manufactured by the Railoc Company of Plainfield, Illinois, for capping grain silos. They fit together with a waterproof seal yet are able to expand and contract with temperature and pressure changes. This makes them ideal for Jantzen's purposes. Just as the ADV is a prototype for a mass-production item suited to what *Domus* magazine called an American "tradition of nomadism with mobile homes," the new Jantzen home is a model for assembly-line modular architecture to house masses of people inex-

FIGURE 357 MICHAEL JANTZEN. Solar Vacation House, Carlyle, Illinois, 1975

This little place is a step on the way to the ADV. Heated by reflective panels and two interior tanks filled with water and with anti-freeze which radiate stored-up heat, the vacation home can be warmed to 100° on a 5° day by the sun alone. The place was designed to be as low cost as possible, using the simplest solar technology and the least expensive material available.

FIGURE 358 Interior of Solar Vacation House

One of the solar windows. Here the solar storage tank just below the window is covered with a padded cushion to form a window seat. Such multipurpose use is typical of Jantzen-designed furniture. Beneath the steps, which lead to the bedroom that forms the second story, is a writing desk that can be leveled for use as a drink tray.

FIGURE 359 MICHAEL JANTZEN. Dome House (in progress), Carlyle, Illinois, 1981

The Jantzen home-in-progress, a design that is intended to be far more efficient than previous structures and will replace the older home, which will become a studio. The new home is a prototype for mass-produced, ecologically responsible dwellings. Made from interlocking steel silo components manufactured by the Railoc Company, the building is very heavily insulated and is indefinitely expansible.

pensively and responsibly. Jantzen's present home—solar and wood fire heated—uses less than three cords of wood during the whole of a severe winter. "If the sun is out," he says, "we don't need any wood. Our house is very efficient. But compared with the dome house I am building, it is inefficient."[4] The dome units, the first of which Mike and Ellen put up in only three hours, are formed of two different size silo roofs, one (24 feet in diameter) set within the other (26 feet in diameter) on a concrete slab, the 12-inch space between them filled with cellulose insulation. The louvered windows were handmade by the Jantzens, but they could be manufactured easily with machine technology. The interior (Fig. 360) has a rather soft, dreamy look despite the materials used, partly because the walls are curved and the sense of being embraced by an articulate space is so palpable. It is typical of photography that spatial effects of interior architectural spaces never come through effectively. Rome's Pantheon seems less spacious in photographs than in reality and so do Fuller's creations and the Jantzen dome house. But the feeling that space is something more than merely distance from barrier to barrier is very real in this house.

The dome house is heated by two sources primarily: the body heat of the inhabitants and the excess heat from major appliances. Cooling is taken care of naturally. During the summer when the first dome went up a prolonged heat wave passed through the region, raising daily temperatures to record heights just below 102°. In the dome, however, the temperature never rose above 78°, even though no blowers or heat exchangers had been installed. The concrete slab itself drew sufficient coolness from the

FIGURE 360 Interior of the Dome House

An interior view of the home-in-progress.

earth to keep the heat down. Unlike geodesic domes, Jantzen's silo domes do not require caulking or gaskets. They are not only mass-producible but have, in fact, been mass-produced successfully for over a quarter century—as silo roofs. The largest of these domes currently available is 32 feet in diameter. Size is the one obvious problem with the Railoc domes, for it is obvious that, unlike Fuller domes, something made from sections of rolled steel has strict limits imposed upon its potential scale since sheets of steel become flimsy after a certain optimal extension. But the limitations are not absolute. Railoc Company has constructed a hemisphere about 100 feet across by using a geodesic dome as a supporting armature or substructure. It is easy to see that it would be possible to construct two-story domes by setting them upon very low cylinders—like squat grain silos—or by going underground and roofing a subterranean home with domes.

As you can see from the reproductions, Jantzen's home-in-progress is not made up of a single dome but is a combination of them. His notions about this are rather intriguing since he sees the potential for creating family dwelling units that not only are autonomous in being independent of utility grids and the like, but also indulge the privacy of various family members. It is even possible that the principles of the Autonomous Dwelling Vehicle might be combined with those of the dome house to permit youngsters to ''dome out,'' as it were, when they come of age. That would be quite in line with old-fashioned American pioneer values. Mike and Ellen Jantzen foresee a future in which a sort of network of technologically superior frontier-type living units will replace the urban centers that exist today. Recently, Jantzen commented:

> As I look at technology and see where it's headed, I think we'll find more people working at home and communicating electronically. People will be working within small communities, developing self-sufficiency. Because of electronics, computer network television, radio, and more sophisticated methods of communication which will be developed, people won't feel isolated in rural areas. As more and more people find they don't need to live in the city, the importance of these autonomous dwellings will increase.[5]

The energy-conscious homes we have surveyed in these few pages are unusual in being individual projects instead of undertakings by heavily funded alternative-energy organizations or giant corporations. They have received some notice in the United States, certainly,[6] but it is in countries like Japan and

Italy, where housing needs are critical and energy costs astronomical, that Jantzen's designs have drawn the most enthusiastic attention. They show, it seems to us, great promise for future directions in mass housing. Prediction is a hazardous business, however. The future does not often work out with the logic hindsight applies to things past, for the unforeseen and neglected seem invariably to have been more important than anything one might have anticipated. Already, it is evident that the very popular and seemingly sensible emphasis on the use of renewable fuels in place of fossil stuffs or nuclear power may turn out to be as inhumane as radiation disease. The main fuel for the Third World—for cooking, heating, and small industry—is cut from forests, and the drive to advance the technology and standards of living of emerging countries has led to the devastation of the woods. This not only threatens the existence of the trees themselves but also menaces the very existence of life. Groundwater, streaming over hillsides denuded of vegetation to hold it back, carries away topsoil, fills rivers to flood stage, and buries dams in silt. Even the dung of animals, which usually fertilizes fields and meadows, is lost, carried away to be used as fuel in place of wood. The consequences of deforestation in Nepal have caused the people to abandon half of the arable land. On the Ivory Coast of Africa, a third of the forest has disappeared in just ten years. The ecological consequences are disastrous, and the chances of bringing a halt to this kind of desolation without running *some* kind of risks seem slight indeed. Still, the route that people like the Jantzens have taken, a direction that involves mainly conserving everything, including the woods, seems to hold out the best chance for humankind.

FIGURE 361 MICHAEL JANTZEN. A table designed to be made from a half sheet of plywood.

In Jantzens' case, the use of sun, wind, and human energy in place of any consumable is fundamental to their philosophical program for conservation, but their example goes far beyond the obvious. Consider Michael's use of wood as a structural material. The table in Fig. 361 has been designed in such a way that it can be made from only half of a single plywood panel. Everything in the houses is conceived with such economy of means in mind, so that it stretches material while conserving energy and draws less from the environment than its owners give back.

All of us quite possibly stand at one of those junctures in history similar to the one dividing Ancient Rome from its Christian future or separating the Middle Ages from the Renaissance. Old conceptions seem outworn in economics, politics, diplomacy, religion, ethics, and even family life. Technology and science are often blamed for this situation but the Jantzen lifestyle is based upon the feeling that "technology isn't something unnatural or something to be afraid of. We just need designers who are enthusiastic about it and know how to *utilize* and *humanize* it."[7]

exercise for chapter 10

EXERCISE 10–1: A Furniture Design

Materials: *White drawing paper*
India ink
Tools: *Drawing instruments*

Design a piece or some pieces of furniture to be fabricated from nothing more than one half sheet of double-faced three-fourths-inch plywood measuring 4′ × 8′. The portion of the sheet used may be a 4-foot square, a 2-foot by 8-foot strip, a triangle formed by cutting a diagonal from corner to corner, or any other form that produces a shape identical with its remainder. Mirror shapes are identical for the purposes of the problem because the "back" and "front" of the plywood are interchangeable. Fasteners, such as nails and screws, and adhesives may be used to hold the structure together, but if you are able to devise a functional item that fits together without such things, that would be all the better.

It may seem incredible that much of anything could be squeezed out of a sheet of wood so small. One of the authors appropriated a lawn/deck chair (Fig. 362) designed to harmonize with Ravine House and revised it slightly to see how much plywood it would have taken to do what was done originally with high-grade redwood planks. We not only had enough wood for the chair (Fig. 363), we had enough left over to build a little coffee table! Granted, the table (Fig. 364) is not quite as large or stable as one might wish for a living room, and we did cheat a bit by putting a glass top on it, but even without glass the table is serviceable as a wall table for plants, magazines, or anything else that is at least 2 1/2 inches wide. Consider,

FIGURE 362
JOHN ADKINS RICHARDSON. Lawn/deck chair. Redwood, 29¹/₂″ tall. 1965.

FIGURE 363 Lawn/deck chair derived from the design in Figure 362. Plywood, 28¹/₄″ tall, 1981

FIGURE 364 Table. Plywood, 17 × 48 × 13″. 1981

though, that you could use a full sheet of plywood to create *two* chairs and have enough lumber left over to create a stronger, wider, and more functional table.

When you lay out the elements of the furniture on a scaled plan like the one in Fig. 365, you must bear in mind that the pieces will probably be cut out by someone like yourself, using a table, radial arm, arbor, or saber saw. Even a cabinetmaker, who would be practiced at making clean cuts part way into a piece of wood, will try to lay out the elements of a job so that the typical cut will be all the way through a piece. In our plan, the first cut would be the one that halves the sheet and the others would follow in the numbered sequence of dotted lines shown on the lower half of the drawing. From the fifth cut on, the work consists of forming individual components from the sections containing them. The isometric blow-up views show the arrangement of the parts.

The actual construction of such a chair is not at all difficult but it will take time and care and will require a few bevel cuts to get a comfortable angle for the seat and back. The Ravine House chairs are, for the most part, held together with nails and white, all-purpose glue. The square, framelike arm-bases on the sides were glued and joined with countersunk woodscrews which were then concealed with dowel pegs. As lawn chairs go these are very durable and comfortable. Made of hardwood and cushioned with leather-covered foam they would be fairly good interior pieces, not unlike Bauhaus versions of the Mission Furniture from the turn of the century. They would go well with the widely available Parsons tables (developed by the Parsons School of Design in the early 1930s).

It should be obvious that our experiment did not produce a terribly economical chair in terms of material use. Were we to have substituted something else for a seat and back—a sling for instance—we could have saved more of the wood and still have had a usable chair. Latticework structures are well-suited to extending materials since they contain a lot of open space, but they have weaknesses such as exposing

FIGURE 365
A: Plan for making a lawn/deck chair and small table from one half sheet of plywood.
B: Blow-up of the chair and table

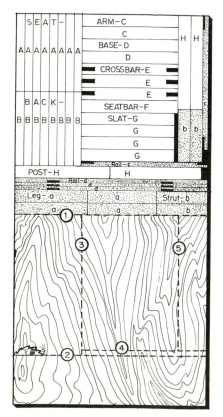

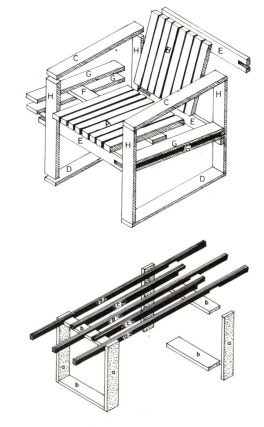

A. The upper half of the sheet shows how the elements were laid out within the space available, with the chair components in white, table elements gray, and waste material black. Sawing directly on the lines would remove between 1/8 and 1/4 inch from the dimensions of the pieces, depending upon the kind of sawblade used, but would not seriously diminish the character of the pieces. All of the parts for the chair are either 2 or 3 inches wide and other dimensions are scaled accordingly. The lower half of the sheet shows how initial cuts would be made in the order numbered.

B. The isometric blow-ups show how the pieces fit together into a chair and table. The seat of the chair is slanted at 5° and the back at 70°, a combination that trial and error proved to be quite comfortable.

many surfaces to damage and offering just as many to be dusted. It would therefore be wise to consider wholly different approaches. Also, remember that chairs and sofas must be strong enough to support heavy people. Tables can be quite flimsy and still hold up a whole meal for as many large eaters as you can seat. Many of us have, at some time, made do with a folding card table for serving dinner.

Fig. 366 is a selection of folding furniture designed by the Jantzens. All of them collapse into themselves in such a way that they either become their own storage compartments or shrink up into easily packaged panels. Conceiving of tables and chairs like these is a good deal more challenging than fitting the parts of a latticework design into a restricted space, but it is also a more worthwhile undertaking.

Undoubtedly, there are many other ways to go about solving the problem embodied in the exercise. Stretch your imagination.

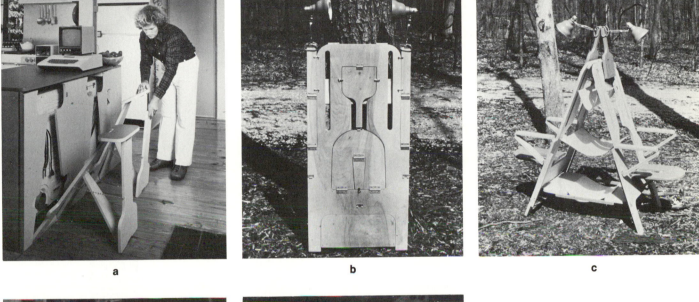

a b c

d e

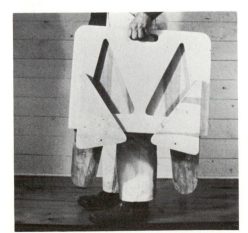

f g

FIGURE 366 MICHAEL JANTZEN.
Folding furniture.
A: Counter seats
B,C: Collapsible reading chairs
D,E: Collapsible sewing center
F,G: Folding chair

237

[1] They are, however, more similar than we seem to believe. To take an obvious example, for a long while it was almost impossible to find a front-wheel drive automobile. Automobiles, houses, and suits in a given price range are practically identical. Of course, there are genuine differences in the quality of competing products. Even so, the differences between an excellent refrigerator and a mediocre one are slight in terms of performance and longevity.

[2] Harry H. Hilberry, "Early Modern Architecture: or, Who Put the Gingerbread on Grandma's Front Porch?" Fine Arts Lecture Series, Southern Illinois University, Edwardsville, June 8, 1966.

[3] Mr. Bakewell estimated that the ADV cost about $16,500 to build, exclusive of his and Jantzen's time.

[4] Quoted in Mildred Arnold, "Art in a Different Form," *Alumnus,* 9, no. 3 (Spring 1981), 19.

[5] *Ibid.*

[6] The most complete article on the Autonomous Dwelling Vehicle is in the April 1980 issue of *Popular Science,* pp. 84–87, 166–171.

[7] Quoted in Arnold, p. 19.

chapter 11

conceptual art and the dematerialization of design

All art, even the most banal, is to some extent conceptual, but very little of it is Conceptual Art. When we look at a traditional painting such as Holbein's *The French Ambassadors* (Fig. 367), we are captivated first by the forms which, for most of us, instantly assume a representational character. We know that the painting is not reality and we are impressed with the skill of the artist; the sheer artifice involved is part of the meaning of the picture. Then, too, it is obvious that there is more to the picture than meets the eye. That is, the strange shape on the floor, apart from anything else, suggests that there is some meaning hiding behind what is immediately evident. There is a whole branch of art history concerned with the symbolic intentions of pictorial imagery; we have earlier drawn upon it in demonstrating that the anamorphic skull on the floor relates the painting to the Vanitas tradition. That resolution of the "puzzle" might seem to have exhausted whatever major secrets the Holbein conceals. But there are other ways of looking at the Holbein. We might, for instance, take a sociological view of it, noting that the globe on the lower shelf charts Magellan's world-circling voyage begun fourteen years earlier with plans for conquest in mind. It is also relevant that highly detailed oil paintings of this kind seem to "possess" whatever objects they portray. It has several times been pointed out that owning a picture of a thing allows you to possess at least the "look" of it and that there is a sort of avid desire for possession revealed in the appearance in history of the kind of superrealism Holbein practices. The stance of the ambassadors is haughty and guarded; they belong "to a class who were convinced that the world was there to furnish their residence in it. In its extreme form this conviction was confirmed by the relations being set up between colonial conquerer and the colonized."[1]

The second analysis gives us a second layer of meaning to apply over the first, and we might also congratulate ourselves on having "demystified" the work by revealing the economic motivations underpinning it, by bringing the artist's aims to account. Holbein's concept of the painting has been exposed and the obsession of society with property has been shown to be so ubiquitous that it penetrates to the very basis of a whole painting style. The analysis strongly supports the Marxian idea that realistic oil painting arose in response to the emergence of capitalism rather than from some simple, straightforward interest in copying the world. To some people this kind of thing seems farfetched. To others, the people who find it so seem very simpleminded. But, whatever the case, it is true that the conceptual nature of *The French Ambassadors* extends beyond the superficial appearance of the work. Even the most conservative interpretation of the importance of the Holbein—the view that it is a sort of sanctified holy relic of aesthetic history and is on a spiritual level removed from such coarse things as commerce and politics—appeals to the notion of a conceptual level beyond that of paint upon canvas.

Politically, Romare Bearden's collage (Fig. 368) is at an opposite pole from Holbein's. It is not concerned with an expression of authority, at least not directly; it expresses the interior viewpoint of a people who have authority imposed upon them, a people who have been enslaved and whose culture is held captive by the ruling majority. Too, although *The Prevelance of Ritual: Baptism* attempts to convey specific meanings, it departs from the conventions of tradi-

FIGURE 367 HANS HOLBEIN. *The French Ambassadors.* 1533. Oil on panel, 81¼ × 82¼″. National Gallery, London

The specific elements represented, as well as the unambiguous, exact way in which they are rendered and the kind of presence the two men present, already imply the characteristics of the period when the ocean routes that would ply the slave trade and provide routes for Western imperialism were being opened up. This implication, along with the Vanitas theme discussed in an earlier chapter, provides a special conception of the image.

FIGURE 368
ROMARE BEARDEN. *The Prevalence of Ritual: Baptism.* 1964. Collage of paper and synthetic polymer on composition board, 9 × 11⅞″. Hirshhorn Museum and Sculpture Garden, Smithsonian Institution, Washington, D.C.

tional painting. Its imagery is unique and is Modern in character, but the meaning is not "visual" in the sense that Ad Reinhardt's painting (Fig. 369) is just what it seems to be and nothing more. Similarly, Philip Hampton (b. 1922) incorporates authentic folk material into designs that are otherwise contemporary-looking (Fig. 370); we have, again, been informed about black history in a way no amount of writing about it can. If one knows something about

FIGURE 369
AD REINHARDT. *Abstract Painting, Blue.* 1952. Oil on canvas, 75 × 28″. Museum of Art, Carnegie Institute, Pittsburgh

FIGURE 370 PHILIP HAMPTON. *I Remember the Core and the Who.* 1982. Mixed media construction, 18 1/2 × 18 1/2″. Private collection. The hand-lettered "family tree" was done by photo-stenciling and a document (c. 1870) owned by Mr. Hampton's great, great aunt and handed down from one generation to the next.

FIGURE 371 FLOYD COLEMAN. *Neo-African #12.* 1977. Mixed media on paper, 9 1/2 × 9 1/2″. Private collection

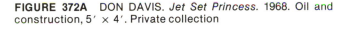

FIGURE 372A DON DAVIS. *Jet Set Princess.* 1968. Oil and construction, 5′ × 4′. Private collection

FIGURE 372B The Alhambra, Granada. 14th century. The Davis painting is derived in part from the pattern beneath the window ledge.

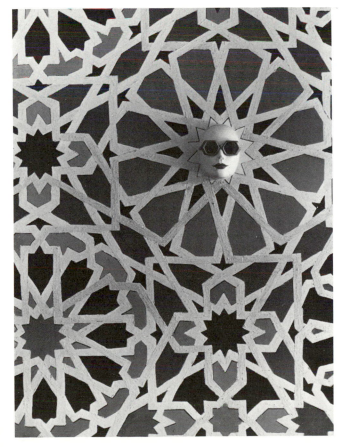

the formal sources of Floyd Coleman's drawing *Neo-African #12* (Fig. 371), it is apparent that its meaning is more comprehensive than any design motif suggests. On a more frivolous note, consider *Jet Set Princess* (Fig. 372A) by Don Davis (b. 1932). This example of Pop Art fuses a design inspired by the ornament of Spain's Alhambra (Fig. 372B) with a blue-lipped mannequin face in sunglasses. It seems to be about

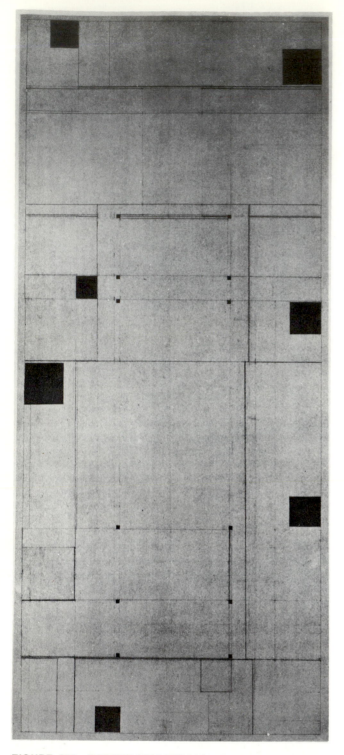

FIGURE 373 ROBERT KIRSCHBAUM. *Temple of Solomon.* (State 2). 1978. Diazo print, 48 × 20″. Private collection

trifles, hardly more meaningful than any travel-theme display in an expensive department store, but the effect of such elaborate tongue-in-*chic* is somehow more ironic than it is derivative. In any event, the meanings of the Bearden, Hampton, Coleman, and Davis are not equivalent to their appearances. Certainly, the same is true of works like those by Kirschbaum (Fig. 373) and Olson (Fig. 374).

Each of the above works is somehow involved with conceptions of art and life in ways that are not at all comparable to Reinhardt's painting, in which the design is the meaning. Reinhardt intended to do nothing more, to convey nothing beyond what he did—a visual display. He once said: ''The one thing to say about art is that it is one thing. Art is art-as-art and everything else is everything else.''[2] A great deal of nonobjective painting and Op Art is similarly insular and, if you wish, ''tautological.'' We have here, then, a broad range of relationships between the appearance (or form) of works of art and the meaning (or content) of them. At one end stands the very materialistic Holbein, at the other end Reinhardt's purely formal design, and, in between, the symbolic montages and emblematic arrangements. Beyond this seemingly complete range, which can be made to comprise architecture and sculpture as well

FIGURE 374 JACK OLSON. *Untitled.* 1981. Acrylic, graphite, prismacolor on canvas, 50 × 60″. Private collection

Of these works, only Ad Reinhardt's painting is equivalent to itself as a visual phenomenon. And even the Reinhardt is an object that is in the possession of a museum; since it is conceivable that it can be purchased at some time, it is a potential element of commerce. As an *object,* it is not unlike the other works.

as two-dimensional media, resides what is an antithetical mode called Conceptual Art.

Conceptual Art is a kind of portmanteau or "catch-all" term used by contemporary artists and critics to describe creative undertakings that are neither "scientific" in the strict sense nor "theatrical" in a traditional way, but which produce art works that are mental conceptions of some sort rather than physical objects such as paintings, sculptures, or buildings. Conceptual Art works may entail the use of pictures, documents, recordings, and so forth, but the work itself is supposed to be a mental synthesis of the material. It might seem to the ordinarily thoughtful person that the kinds of concepts embodied in the Holbein painting might, after all, correspond to such a synthesis. Most Conceptual Art workers would not think so, because their principles are on the order of a kind of overarching, all-encompassing Post-Modernism and the pretensions of traditional art are irrelevant to their concerns. That is to say, theorists of Conceptual Art might be interested in the Holbein as an historical fact, might admire the skill required to create it, might even take pleasure in the visual niceties of the composition, but few of them would find it very closely connected to what they mean when they speak of Art. In fact, many Conceptualists would not consider the contemporary works very relevant either. Joseph Kosuth (b. 1945), a germinal thinker prominent among them, has asserted that "art's viability is not connected to the presentation of visual or other kinds of experience."[3] To anyone unfamiliar with the history of contemporary art and art criticism, a statement of this kind sounds positively incredible. Nonetheless, it is a very prominent attitude expressed by a number of serious people in the art world today and it has considerable influence in the field of art and design.

Whatever else one may think of Conceptual Art, the activities of those who disavow art objects as such have been salutary for creative expression, especially by way of cooperative ventures by avant-garde workers in such disparate disciplines as music, theatre, dance, cinema, and electronics. As it happens, however, most of what these people do is without much visual interest. It also usually involves sociological concepts of the shallowest possible order and ideals of human behavior that are dreary beyond all comprehension—largely because they are so readily grasped, can be taken for granted, or are absurd. The exceptions, though, are of considerable interest. As the basis of any defense of the form invariably we find the notion that the most important thing about True Art is ideational, at last, and that the material properties of a painting or a sculpture serve to convey something that is ultimately nonsensual, or, conversely, tend to misdirect attention from the art in the art work to its value as an object of commerce. It should be noted that Conceptualists are frequently very antagonistic to the idea that visible aesthetic appeal has anything to do with art as opposed to entertainment or decoration and also that their art works have often expressed a very distinct social ideology. Since ordinary artists produce things that can be purchased, their art easily fits into the value scheme of routine commercial activity. By eliminating the possibility of anyone's actually possessing their art, some Conceptualists hope to overcome the prevailing commercialism of the art world.

Like many things in the arts, Conceptual Art is more easily defined ostensively—that is, by the giving of examples—than it is by words and propositions. Of course, people will not always agree on whether a given example is *really* an example. That is typical of normal conversation; what one of us calls a "stick" another will refer to as a "board" or "sliver." Some of the people we are about to use as examples would not be considered even vaguely within the realm of Conceptualism by the "purer" Conceptualists, whose work tends to be purely ideational. For that matter, the authors do not agree among themselves on all of the selections.

Among artists sometimes celebrated as a Conceptualist is Christo (b. 1935 as Christo Javacheff). In 1969 he draped a mile of the Australian coast in one million square feet of polyethylene plastic. His *Running Fence* (Fig. 375), which for two weeks in 1976

FIGURE 375 CHRISTO. *Running Fence.* 1976. Sonoma and Marin Counties, California. Nylon fabric, steel poles, and cables, 18′ tall, 24 1/2 miles long. Photo: Wolfgang Volz.

This art work is a physical object 24 1/2 miles long that has been photographed. It required for its construction all kinds of people involved on many levels but it no longer exists. It was conceived of as undertaking more than as a structure.

stretched 24 and a half miles across Sonoma and Marin Counties in California, was another undertaking that challenged the artistic eliteness of avant-garde design. Christo's projects are normally of such vastness and normally entail the granting of so many easements, the securing of so many foundation grants, and the cooperation of so many assistants that they constitute communal activities. The participants who actually construct the pieces are, for the most part, students, artists, and various other kinds of enthusiasts, but Christo's undertakings also draw into their orbits bankers, local, state, and federal government officials, and foundations representatives. For it takes all sorts of people working together to make possible such impressive projects. The *Running Fence* ornamented the rolling hills of two wealthy suburban counties in northern California and in two weeks it was gone, a collective enterprise that vanished without so much as a trace of impact on the environment. Whether such an expenditure of time and human labor was justified by so transient a project is something that must be left to sheer opinion. The fact that it was completed, then deleted, with no negative effect upon the land is perhaps itself a demonstration of human responsibility worth undertaking.

Walter de Maria (b. 1935), who created *Lightning Field* (Fig. 376) 40 miles outside of Flagstaff, Arizona, is intrigued by invisibility and transient im-

FIGURE 376
WALTER DE MARIA. *Lightning Field.* 1977. © Dia Foundation. Photos: John Cliett.

The array of lightning rods in a field 40 miles from Flagstaff has the properties of faintness in the distance and also a physical presence that produces a dance of lightning bolts during thunderstorms, but it is the concept of creating a delicate pattern that connects the heavens to the Earth that is the real work.

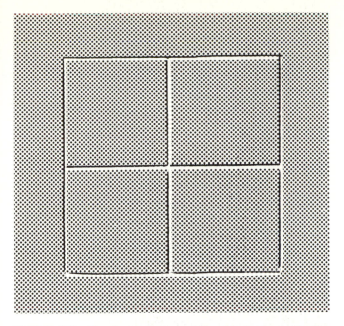

FIGURE 377 The design that would have been formed by the superpositioning of Walter de Maria's *Three Continents* earthworks.

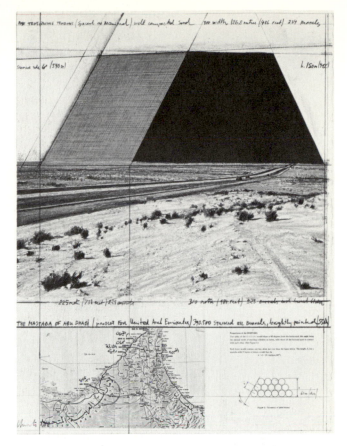

FIGURE 378 CHRISTO. *The Mastaba of Abu Dhabi, Project for the United Arab Emirates.* 1980. Collage, 27 1/4 × 20 1/2″. Photo: eeva-Inkeri. © Copyright: C.V.J. Corp. and Christo.

This proposal features a construction of 390,500 oil barrels, brightly painted and stacked 492 feet high, covering an area 738 feet by 984 feet. (A mastaba is the principal form of ancient Egyptian tomb.)

agery. One must walk the last mile to the *Lightning Field* and the lightning rods are at first unseen. As you approach them, however, they begin to form, mirage-like, from the air of the desert. In thunderstorms, of course, they attract bolts from the great engine of the sky; the artist has then become a sort of choreographer of one of nature's most terrifying forces. Earlier, de Maria had an idea which is more clearly on the order of a pure Conceptualism. In his proposal *Three Continents* he sought support for an undertaking in which he would bulldoze huge surface marks in the deserts of Africa, India, and North America. In the Sahara the mark would run East and West, in India it would be oriented North to South, and outside of Las Vegas, Nevada, he was to scrape off a gigantic, oriented square. These scars fell on the route of an orbiting weather satellite which would photograph them and superimpose them into a design like that in Fig. 377, thus "linking" the continents. He actually got as far as bulldozing a trench for two weeks in the Sahara desert before hostile natives drove him away.

Christo (Fig. 378), de Maria, and Michael Heizer (b. 1944) are Conceptualists whose work might also be identified as *Environmental Art*. Heizer, who is a successful painter, became interested in earthworks and undertook some enormously large-scale "drawings" in 1968, drawings done with motorcycle tire tracks on a dry lake bed. Since then he has done some vast excavations. In the middle of the 1970s he purchased a tract of land north of Las Vegas to construct a three-mile square statement in earth and concrete entitled *Complex of the City*. Of art works one might think of as being Conceptual, this metaphorical comment on architectural space and sculptural form is about as completely materialistic as one can imagine. Most Conceptual Art is less firmly tied to the earth.

Leila Daw (b. 1940) has been doing works involving the use of sky-writing planes which trace patterns in the sky to match the contours of rivers, landscape topography, and so forth, even flying straight down the Mississippi's course so that thermal variations resulting from the different reflective properties of water, land, and foliage create variations in the smoke trail hanging above the landscape. Fig. 379 is a visualization of one of her proposals.[4]

Michael Jantzen, too, enlists mechanisms in his Conceptual Art. For instance, he is fascinated by the possibility of coordinating solar power cells with a buried, pressure-sensitive security alarm system. He conceives of an open field in which the only visible presence would be a pole, or poles, supporting a

FIGURE 379 LEILA DAW.
Partial Documentation of *Prehistoric River Channel*. A portion of the documentation of one of a continuing series of environmentally interactive drawing performances. 1981.

Using an agricultural airplane modified to emit smoke over a great distance, a diagram was drawn in the sky, mapping the probable course of the Mississippi River as it would have been in prehistoric times when today's Horseshoe Lake was a bend in the main river channel that bypassed what is now the city of St. Louis, Missouri. First, the plane traced the prehistoric west bank for twenty miles from north to south and then followed the contour of the east bank for the same distance in the opposite direction. Ms. Daw explains that the bluffs, not the levees, are the true boundaries of the great midcontinental rivers and that today's banks are only temporary designations of the water's edge, "existing more definitely in our minds and maps than in nature. The entire floodplain between the bluffs is the true river bed; every part of it has been under water at some time in the geologic past." Thus, her skywritten diagram not only projects the geography of the river valley backward in time and upward in space, but also shifts and fades away in the winds aloft in a way that is somewhat analogous to the geological changes that have occurred over thousands of years.

loudspeaker and a photovoltaic panel. The panel would supply energy to the buried cable. On sunny days the cable would be fully charged and operate as intended by the manufacturers, having a fairly extensive "sensing" distance and high reliability all along the guarded perimeter. On cloudy days the power would be low and the cable range weak and possibly intermittent. A viewer/participant approaching the pole across the field would perhaps hear an alarm sound announcing that he or she had passed into the ambit of the buried cable. On a sunny day, walking along the line of the cable (which might be laid out in a rectangle but would more likely be given a less pre-dictable configuration) would produce a steady wail from the speaker. On an overcast day the signal would be interrupted and uncertain. In either instance people, in walking on the field, would form a conception of the buried design.

Another Jantzen Conceptual Art work involves four transparent slides made of the four sides of an object or space—let us say of a small building—and then projected at right angles and across from one another over a vacant space, focused on the place where the object might actually have stood (Fig. 380). The only way that a viewer could comprehend the projected image would be by walking into it alone or with others and observing the fragments of the image reflecting from flesh, hair, and clothing interrupting the projector beams. You might, for instance, see a section of clapboard siding making stripes across your chest and see the edge of a windowframe peeking over the shoulder from your back. You've become the screen for the obverse sides of the building.

Jantzen says that this kind of game helps him to realize architectural space differently from the way in which society has conditioned him to preconceive it. However, it is more than just an adjunct to inspiration; Jantzen considers it as artistic as his sculpture or architectural designs. That may be a controversial point of view in some quarters, but Conceptual Art, in its many forms, has in fact become quite fashionable among people highly active in the world of art and design. Ironically, in light of its frequent antagonism to the world of commerce, Conceptual Art has become rather closely tied to the broad field of design because the skills and some of the tools involved overlap those of architects and graphic designers.

Clayton Lee (b. 1951), of the Art Institute of Chicago, composed a whimsical exercise in logic dealing with the matter of the associations among art, fashion, and perversity of taste (Fig. 381). If you work it all through, it will become evident that the possibilities are "branched" so that a decision can come out either of two ways. The formulation of the problem is such that it requires its readers to be consistent and willing to "live" with the consequences of a perhaps arbitrary conviction. Lee exhibited this exercise along with his photographs as another work of graphic art in an art exhibition. Not every visitor to the gallery thought that it should have been included as "art." But then, it addresses itself to that very question. And it brings into question the legitimacy of more traditional works, other exhibitions, and even the criticism of art work. It is, in other words, a provocation which, though done with a sort of forked-tongue-in-both-cheeks approach to things, raises genuine questions about art and art criticism. If it helps us to rethink our cultural circumstances and role as designers, then it is useful to contemplate.

Such modes of Conceptual Art are similar to the studies (Fig. 382) by Conceptual architect Peter Eisenman in getting at the root relationships of connected things we don't always associate with one another. Giving close attention to contingencies and counterparts is a fundamental aspect of designing for a world in which imbalances may be horrifying in their consequences. Those who create works of art, and particularly those who create designs intended to convey explicit messages to the community or serve the human need for shelter, must be ever conscious of the ways in which society frames its problems. Some-

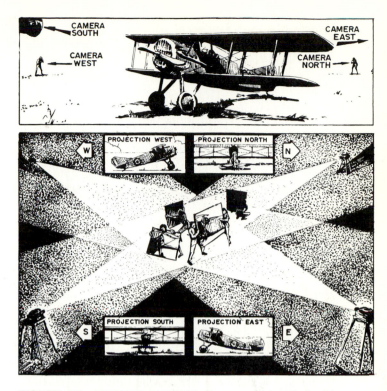

FIGURE 380 MICHAEL JANTZEN. *Conceptual Art Work with Four Projectors.* 1982.

In our artist's visualization, photographs of an antique airplane, taken from four directly opposed viewpoints, are projected into open space from similarly opposed positions while individuals with portable panels act as screens in the zone where all of the images intersect.

times a solution to a problem cannot be found without breaking the frame. That occurs, of course, when the social scheme is itself the fundamental problem.

Every balanced liberal and responsible conservative realizes that there is grave peril whenever a government imposes its preferences upon the cultural life of a people and curtails the freedom of citizens to differ in their opinions, faiths, and taste. However, it is also true that the spread of individualistic attitudes, to which authority is inevitably an object of suspicion and fear, promotes the proliferation of conflicting, self-interested factions which are inimical to the bonds of common purpose in society. It is almost certainly foolish to believe, as many designers seem to have believed, that art can serve as an antidote to social disorganization or that it can be a spiritual substitute for political freedom or economic security.[5] However, it is not so silly to think that designers have a role to play in structuring the frame of human existence so that it is not poisoned by waste, made hideous either by garish conflict or numbing monotony, or depleted by wanton disregard of the environment.

A DECISION MAKING RULE

The decision to be made is whether to assertively prevent full participation in the artworld, assertively participate in the artworld, or to remain inactive.

Given the definitions, assume that the rewards for artistic activity are based on what is best for the public [1].

However, closer examination reveals that there are private as well as public benefits from success. A more descriptive decision making rule could be [2]. Verbally, the probability of an individual being an avantgardist is the sum of the public and private rewards given success (Ls + Li is the probability of success if i is an avantgardist), minus the cost of a permanently damaged career (1 - Ls - Li is the probability of failure if i is an avantgardist), minus the cost of failure given the probability of failure plus the pleasure of participation in the artworld.

Analogously, an individual decides to combat the avantgardists by [3]. The probability of failure and the cost of failure is assumed the same for avantgardists as well as for anti-avantgardists as is the joy of participation.

If Pa = Pz, inaction results, if Pa is greater than Pz then the avantgardists prevail, if Pz is greater than Pa, then the anti-avantgardists prevail. By comparing [2] and [3], we have a decision making rule. This is shown as [4], which becomes simplified to [5].

In an active artworld, the contribution of any individual to success (Li) is very small and (with the exception of "heroines", "heroes", or "leaders") approaches zero. Allowing this we attain [6], which becomes simplified to [7].

Define:
Pi as the probability an individual is an avantgardist or an anti-avantgardist
Pa = avantgardist
Pz = anti-avantgardist
Bg = public benefit of success, e.g. History
Ls = probability of success without individual's participation in the artworld
Li = individual's contribution to the likelihood of success
Ri = individual's personal rewards if her/his side succeeds or wins
Ci = cost of permanently damaged career
F = probability of failure
I = cost of failure
E = aesthetic pleasure of being avantgardist or anti-avantgardist

[1] $Pi = (Bg)(Ls)$

[2] $Pa = Bg(Ls + Li) + Ra(Ls + Li) - Ca(1 - Ls - Li) - (F)(I) + E$

[3] $Pz = Bg(Ls + Li) + Rz(1 - Ls - Li) - Cz(Ls - Li) - (F)(I) + E$

[4] $Bg(Ls + Li) + Ra(Ls + Li) - Ca(1 - Ls - Li) - (F)(I) + E = Bg(Ls + Li) + Rz(1 - Ls - Li) - Cz(Ls - Li) - (F)(I) + E$

[5] $Bg(Ls + Li) + Ra(Ls + Li) - Ca(1 - Ls - Li) = Bg(Ls + Li) + Rz(1 - Ls - Li) - Cz(Ls - Li)$

[6] $Bg(Ls) + Ra(Ls) - Ca(1 - Ls) = Bg(Ls) + Rz(1 - Ls) - Cz(Ls)$

[7] $Ra(Ls) - Ca(1 - Ls) = Rz(1 - Ls) - Cz(Ls)$

Thus, while individual behavior in this model becomes based on private rewards or costs and their likelihood of success, the principal constraint centers on the historical conventions defining the artworld, the public, and their criteria for success.

FIGURE 381 CLAYTON LEE. *A Decision Making Rule.* Conceptual Piece, 1981.

A self-explanatory exercise in logic as applied to art and nonart. It was itself displayed as a work of art, thus challenging the exhibition and, as well, all of the other things displayed within the *art* exhibition.

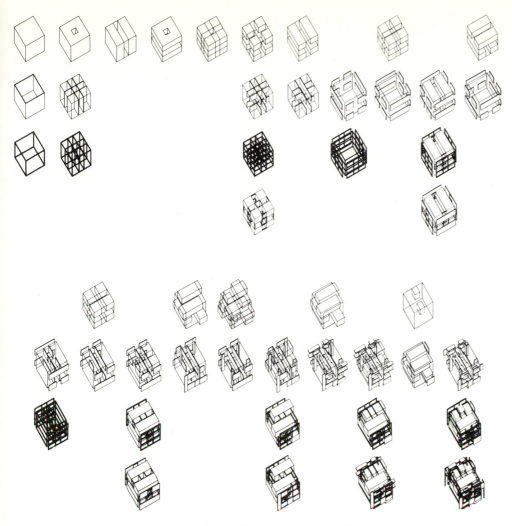

FIGURE 382 PETER EISENMAN. *House IV.* 1971. Ink and white paint on clear mylar film, 51 × 17″. (Delineator, Robert Cole, 1975)

A piece of analysis from a Conceptual Architect.

exercises for chapter 11

EXERCISE 11-1: Conceptual Art Work

> *Materials:* *Whatever you can think of*
> *Tools:* *Anything you can use*

Conceive something nonphysical that will affect the observer.

EXERCISE 11-2: Futurity

> *Materials:* *All*
> *Tools:* *Self*

Think long, hard thoughts.

notes for chapter 11

[1] John Berger, *Ways of Seeing* (New York: Viking, 1973), p. 96.

[2] Ad Reinhardt quoted in *Readings in American Art,* ed. Barbara Rose (New York: Praeger, 1975), p. 138.

[3] Joseph Kosuth, ''Art After Philosophy, I and

II," *Idea Art,* ed. Gregory Batcock (New York, 1973), p. 94. Kosuth's argument, though specifically made in reference to Hegel and twentieth-century philosopher A. J. Ayer, is essentially the same one Plato (427?–347? B.C.) enunciated centuries ago. But if art objects, as opposed to art ideas, have no intrinsic value or significance, why not dispense with artistry altogether? Why bother with even a Conceptual Art? Kosuth argues that the only thing that distinguishes art from logic, mathematics, and science is its uselessness. This line of thought leads directly to the conclusion that the truest Conceptualists are people like mathematician Kurt Gödel, whose concerns existed as pure conceptualization having neither the capacity for application nor even the utility of an example. The theoretical sciences are full of "constructs" which defy all attempts to visualize them, let alone objectify them in material form.

⁴A charming, although unconnected precedent for Daw's sky-writing venture was a cinematic project undertaken about fifteen years ago by humorist-composer-entertainer Mason Williams, best known for his guitar piece, *Classical Gas.* He took a camera crew into the Mojave desert near dawn. At sunrise, sky-writing pilot V. E. Noble received Williams's radio signal and began tracing an outline drawing of the world's largest "sunflower." Unfortunately, the sun was so bright that it blotted out the film image. But Williams wasn't too concerned; he said that seeing it wasn't as important to its meaning as having people hear about it. His view of such activities is summed up in a little four-line poem that closes with the insight that life's "the perfect thing/to pass the time away."

⁵This tendency is particularly evident among Bauhaus-connected theorists. See Richardson, *Modern Art and Scientific Thought,* pp. 156–170. For more general treatments of modernist social affectations see Eugene Goodheart, *Culture and the Radical Conscience* (Cambridge, 1973), pp. 125–162; and Daniel Bell, *Cultural Contradictions of Capitalism* (New York, 1976), *passim.*

glossary

The following list contains definitions of all terms italicized in the book (other than foreign words, titles, and things obviously being stressed). It also contains many terms not used in the text. Although one may expect to encounter these in connection with design, we have not had occasion or space to touch upon them. Cross-references are indicated by words in italics.

abstract art　Art dependent on the idea that artistic values reside in forms and colors independent of subject matter. An abstract work may resemble something else (apples, nude women, etc.), but the stress will be on form and color. Any work can be looked upon as an abstraction, but some styles (*Cubism, Abstract Expressionism, Neoplasticism*) intentionally emphasize the formal over any symbolic value.

abstract expressionism　A style of painting in which the artist expresses feeling spontaneously and without reference to any representation of physical reality. Normally, the term signifies a movement that originated in America during the late 1940s, but Wassily Kandinsky had pioneered the essentials of the manner before World War I.

abstraction　A work that is deliberately *abstract art,* as in: "This painting by Picasso is an abstraction."

Academy, Academicism, Academician　In the fine arts, an Academy is an officially recognized school of professional art instruction, or its equivalent in terms of prestige and authority. In particular, the term refers to the French Academy of Painting and Sculpture and its imitators in other countries, which became centers of such opposition to new styles during the nineteenth century that the term "Academic" is now synonymous with pretentious dullness. This view is not justified by the facts, but it is true that the Academies tended to emphasize conventional subject matter, *classicist* form, and the technical mastery of very specific skills (and not others) and tended to use "finish" as an ultimate criterion of quality, much to the detriment of modes like those of *Realism* and *Impressionism,* which emphasized directness, candor, and spontaneity. To an Academician, a painting by Claude Monet seemed, at best, a delightfully colored sketch rather than a completed painting; to an Impressionist, the Academic prize winner was overworked and artificial. Actually, both of these points of view were correct, given the assumptions that prompted them, but the modernists won the contest for the taste of the future and it is their view that has predominated since the early twentieth century.

action painting　A type of *Abstract Expressionism* incorporating impulsive gestures.

additive color　See Index.

aesthetic　Having to do with art and beauty. Often used synonymously with a theory of art or to refer to the characteristics of a given style, as, for example, "the Impressionist aesthetic." (Sometimes spelled *esthetic.*)

agate　A unit of measure used to calculate column space in newspapers. One inch equals 14 agate lines.

airbrush　A small hand-held spray gun used by graphic artists for retouching photographs and for producing *continuous tone* illustrations.

alley　Denotes the horizontal and vertical spaces which separate columns or blocks of type from one another in the way *margins* separate type from the edges of a page.

anamorphic A form of regularized distortion of an image so that it looks correct only when seen from a certain angle or when viewed with some special device. See Index for illustrations.

arc lamp The light source used in making *plates*. Usually, it is produced by a current arcing across two carbon electrodes.

Armory Show The first large, public exhibition of modern art in the United States. Organized by *The Eight* and some sympathizers, it was held in the 69th Regiment Armory building in New York City in 1913.

art decco A design vernacular of the 1920s, inspired by *Art Nouveau,* but incorporating various modernist mannerisms such as *Cubism, Futurism,* and *Expressionism.* The greatest monument to the style is William van Alen's Chrysler Building (1926–1930).

Art Nouveau A decorative style of the 1890s which attempted to break with past traditions. It is characterized by interlacing plant forms and similar organic treatments along with relatively flat color treatments.

Artype Brand name for a brand of cut-out *transfer type.*

ascender See Fig. 291.

atmospheric perspective The illusion of depth in painting created by reduction of contrast between lights and darks, cooling of colors, and blurring of outlines as things represented recede from the picture plane.

avant-garde French word for "vanguard." Artists who are unorthodox in their approach. The connotation is that such people are "ahead of their time," but it sometimes happens that they are merely eccentric. Still, *Realism, Expressionism, Dadaism, Cubism,* and the others are all examples of avant-garde movements.

axonometric projection A form of engineering drawing in which all surfaces of the object represented are at an angle to the picture plane. See Figure 227 for an illustration.

backbone The spine of a book—the part connecting the back and front covers.

backing-up Printing the reverse side of a sheet.

baroque Generally, the seventeenth century. The term connotes grandiose elaboration in architecture and decoration, since that is generally typical of the period. However, in France the stress was on monumental dignity, and Dutch art of the period is so different that it has been distinguished by the name "Protestant Baroque."

basis weight The weight in pounds of a *ream* of paper that has been cut to a standard size called "basis size," a size that varies with different grades of paper.

bastard size Anything nonstandard in size.

batik A process of dyeing cloth by painting designs on it in wax so that only exposed areas are impregnated with dye. The wax is removed after the dyeing.

Bauhaus A school of industrial design founded in Germany in 1919 by architect Walter Gropius. Known for its attempts to integrate art and technology.

benday pattern In commercial printing, the technique invented by Benjamin Day (1838–1916) which adds tints to *line cuts* by means of dots, lines, and other regular patterns.

bite Narrowly, to etch a piece of metal. Loosely, in *etching* and *photoengraving* this is the time needed to *etch* a given depth into the metal.

Blaue Reiter, Der "The Blue Rider." A branch of German Expressionism centered in Munich. Tended to be far more *abstract* than *Die Brücke.*

bleed To print a design onto paper so that its limit extends beyond the edge that will be trimmed. In other words, to eliminate one or more margins.

blocking In *letterpress* printing, this is mounting a photoengraving onto a block of wood so that it will be raised to type height.

blowup An enlargement of *copy.*

blow up To enlarge *copy.*

blow-up view A form of technical illustration which represents the individual parts of a thing separately and apart so that an observer can see the way they fit together. Sometimes called an "exploded view."

bold In typography, a heavy version of a typeface.

book paper A category of paper used for printing. The other categories generally used are *pulp paper* and *newsprint.*

bozzetto See *maquette.*

brayer A hand roller for applying printers' ink to surfaces.

bristol board Heavy, opaque commercial drawing papers. Most commercial illustration and cartooning is done on this kind of paper. It comes in three weights—one-, two-, and three-ply—and in a slick, "plate" finish or a matte, "kid" finish. Even three-ply is not well-described by the word "board," but kid bristol *is* the paper used to provide the surface for "illustration board."

broken color The technique of applying paint in brief, rather heavy strokes over a background color so that the latter shows through in patches. Typical of *Impressionism.*

Brücke, Die "The Bridge." A branch of German *Expressionism* centered in Dresden. Quite similar in appearance to *Fauvism* but more pessimistic about things in general.

built-up lettering Letter forms that have been built up from a number of individual strokes, often drawn mechanically.

cabinet projection A form of *oblique projection.*

camera ready Describes *copy* suitable for photographing by a process camera to produce negatives to be used in platemaking.

caption Text accompanying illustrations.

cartoon A full-size preliminary *drawing* for a painting. Also, a frivolous or satirical drawing. (The former is the original meaning but is retained today only in art circles.)

cavalier projection A form of *oblique projection.*

central projection Technically correct name for scientific *perspective.*

ceramics Pottery making. More specifically, objects made of clay and fired at high temperatures to render them stronger or waterproof.

characters All of the individual letters, numerals, and punctuation marks in a given typeface.

charcoal A drawing medium produced by charring organic substances (wood or bone) until they are reduced to carbon. A drawing made with charcoal.

chase In *letterpress* printing, the frame into which type and *engravings* are clamped or *locked up* for printing. In metalsmithing, ''to chase'' is to ornament metal by indenting it with a hammer and tools that do not cut it. See also *repoussé.*

chiaroscuro Light-dark relationships in a work of art. See Index.

chip A sample of a color or, occasionally, a texture.

chipboard A substantial paperboard made from wastepaper. Gray in color. It is used as a support for cloth or padded covers in bookbinding and is ideal to use as backing for *mats,* as a cutting surface, and for packing.

Classical Narrowly, the term refers to the art of the ancient Greeks during the *Golden Age* of the fifth century B.C. By extension it is applied to all the works of antiquity from 600 B.C. through the fall of Rome. By still further extension, Classical is used to describe any form thought to be derived from Greek and Roman examples; thus, the art of the Italian *Renaissance* is sometimes called Classical. It is, however, rather more common to refer to styles derived from the Antique as being *Classicist.* To be Classical implies perfection of form, emphasis on harmony and proportion, and restraint of emotion. Normally, the term is applied by art critics only to art that is idealistic and representational. Of course, it is commonplace today to describe something that is an outstanding representative of its type as being a ''classic.''

Classicism See *Classical* and *Neoclassicism.*

coated paper Paper whose surface has been covered with pigment and an adhesive to improve the printing. Usually thought of as being glossy *stock,* it may be dull as well. It is especially well-suited to *halftones.*

collage A work made up in whole or in part by gluing various materials (newspaper, wallpaper, cloth, photographs, bits of wood, etc.) to a piece of canvas or other *ground.*

collagraphy A *relief printing* method which uses as the printing surface cardboard shapes and pieces of materials glued onto a base. It is often used in conjunction with *intaglio* or *lithographic* techniques.

collate To arrange pages in such an order that when they are bound they will be read in the correct sequence.

collotype A printing technique using gelatine-coated, light-sensitive *plates.* Good for extremely fine and precise color reproduction of very limited *press runs,* it produces truly graduated values without any *halftone* dots.

color terminology Printers use slightly different terms for color characteristics than do artists or laypersons. What most lay people call ''the color,'' artists call *hue* and the printer refers to as ''chroma.'' Darkness or lightness, which artists usually term *tone* or *value* is called ''gray.'' Intensity of hue—how bright or dull the color is—is named *strength* by printers.

column inch A measure used by small newspapers. It is based on a unit of space one column wide and 1 inch long.

combination plate See Index.

complementary colors Colors that are opposite each other on a colorwheel. When mixed together in proper proportions they form a neutral gray.

comprehensive An accurate representation of a design as it will appear when printed. In effect, it is an artist's conception of exactly what effect a design will have, containing all colors, display lettering, and so on. Commonly, a comprehensive is referred to as a ''comp.''

Conceptual Art See Index.

condensed In typography, a modification of a typeface, the characters being narrower than the standard version. The opposite of *expanded.*

connected dot In either a *plate* or *negative, halftone* dots that touch one another. Thus, the dot pattern formed in a halftone of 50 percent or greater value.

connoisseur Technically, an expert on art whose profession is identifying the specific artist who created a specific work, determining the age of a work, or making similar attributions. Generally, a person of highly developed artistic sensibilities or an aficionado of some kinds of artifacts.

Constructivism   A twentieth-century movement in sculpture which emphasized precision, technology, and *nonrepresentational* form.

continuous tone copy   *Copy* containing genuine grays as well as black and white. Usually a photograph but also art work that has the same effect.

cool color   Blue and such hues as approach blue.

copy   Designates all artwork to be printed. Also refers to typewritten material to be typeset or otherwise prepared for printing.

copyright   The legal control of reproduction of a work.

crop marks   Short lines drawn on an *overlay* or on the margins of a photograph so that if they were all connected they would indicate where a picture is to be trimmed.

cross-hatching   The production of relative *value* relationships in drawing with sets of more or less parallel lines placed on top of one another at varying angles.

counter   See Fig. 291.

counterchange   See Index.

Counter-Reformation   The reform of the Catholic Church during the sixteenth century. It had significant effects upon all art produced under the patronage of the Church. (Actually, the reforms had begun before the rise of Protestantism, and Catholics therefore prefer the term "Catholic Reform.")

Cubism   The first *abstract* style. Invented in 1909 by Pablo Picasso and Georges Braque, it revolutionized painting by stressing formal relationships over representationalism. It has been falsely identified with the representation of multiple viewpoints of objects because of a few isolated examples and an "explanation" put forth by the minor Cubists, Metzinger and Gleizes. See Index for examples and for further commentary.

cursive   Lettering that has the letters in each word connected to one another by strokes. Also, typefaces that resemble handwriting, even when the letters are not connected.

cut   Any printing plate. In Europe the term in English is "block."

cyan   Process blue. Turquoise.

Dadaism   A *movement* begun in Zurich during World War I. It reacted to the war by expressing the absurdity of all conventions and the futility of all acts. It soon became international in scope. Its attitudes and examples have remained influential up to the present moment.

day-glo   See *fluorescent ink.*

dead metal   Areas in a *photoengraving* that must not print and are routed out.

deep etch   When used in connection with *photoengraving* the term refers to *bites* beyond the initial etch. In *offset lithography* the reference is to recessing of the plate below the surface of the printing area in order to use it for very long *press runs.*

delineator   An artist whose specialty is doing projective renderings of objects, usually buildings, from *orthographic projections* and detailed information.

depth of field   The region beyond the camera lens in which things are in focus. The greater the distance between the nearest and farthest objects that appear sharply focused, the greater the depth of field. The smaller the aperture (lens opening), the deeper this field.

descender   See Fig. 291.

designer colors   See Index.

die-cut   To excise a portion of a sheet of paper or cardboard by striking it with a set of steel blades formed around a die.

dimension marks   Arrows drawn on the margin of a *mechanical* beyond the area to be reproduced to indicate the size of the printed image. In the break between the two arrows a notation gives the amount of reduction. (See Fig. 109.)

dimetric projection   A form of *axonometric projection.*

diptych   A two-panel picture, normally hinged so that it can be closed like a book.

display type   Special, decorative alphabets in type form, seldom used for anything except headings, titles, and so on.

distemper   Any of a number of water-base paints using a simple glue or casein as the binder. Calcimine, show-card, and poster colors are examples. (The "tempera" of the primary grades is a distemper, not a *tempera;* the confusion arises because of a trade name for one brand, Tempra.) The term is more commonly used in Britain than in the United States.

double burn   Exposure of a *plate* to two or more *negatives.*

drawing   The projection of an image on a surface by some instrument capable of making a mark. Most often, drawings serve as studies or sketches for a work in some other *medium,* but they are frequently done as completed works in themselves.

drop out   To produce areas of solid white in a *halftone,* as is commonly done with titles and other lettering in ads, in magazines, and on book jackets.

dry-brush   To draw with a brush—usually a small red sable brush—and India ink. A very common technique in commercial illustration. (Fig. 120B is a dry-brush drawing.) Compare *wash drawing*.

drypoint The simplest form of metal *engraving*. Done by scratching a soft metal—usually copper—with a sharp steel or diamond-tipped needle. Sometimes called "drypoint etching," although no *etching* is involved.

dummy The mockup of a printed piece, showing where all the elements are to fit.

duotone A two-color *halftone* made from a single black and white photograph. Two *plates* are made, one emphasizing extreme contrasts of *value*, the other emphasizing middle values. The first plate is run in black ink over the second, which has been printed in color. This creates a printed image with a full range of values and the appearance of being a rich *monochromatic* variation of the values of the color plate. Compare *fake duotone* and *flat-tint halftone*.

earthworks A form of sculpture in which the artist designs a pattern that is carved, dug, built, or scraped into the landscape itself.

editing Checking *copy* for quality, consistency, grammatical correctness, style, and other things before it is released to the printer.

eggshell paper Drawing or printing paper having a surface similar to the texture and color of the shell of a hen's egg.

Eight, the A group of painters active in New York City at the beginning of the twentieth century. For the most part they were *realists* with strong social convictions, and they took their subjects from the streets and back alleys, the tenements and their immigrant inhabitants. They are better known today as the Ash Can School, an appellation that most of us suppose was tacked onto them by critics offended by the raw candor of their work but that actually was first applied openly in 1937 by Helen Appleton Read in her introduction to the catalogue for the exhibition, *New York Realists, 1900–1914* at the Whitney Museum of American Art. The term seems to have been invented by cartoonist Art Young, a doctrinaire socialist, who told the leader of the Eight, John Sloan, that the only "revolutionary content" in their works were the ash cans. (See William Innes Homer, *Robert Henri and his Circle,* Ithaca: Cornell University Press, 1969, pp. 130 and 230, n. 18.)

elevation One of the "views" of an object in *orthographic projection*.

elite The smallest regular typewriter type. It has 12 letters per inch, compared with 10 in *pica* type. See also *pitch*.

emboss To raise up from a surface. Characteristic of line *engraving*.

encaustic A painting *medium* using hot, colored waxes. It is extremely *permanent*.

engraving The process of cutting a design into a substance, usually metal, with a sharp tool. Also refers to a print made from an engraved *plate*. Generally used in the commercial printing trade to mean something quite different, a *photoengraving*.

environmental art Art work that becomes part of the landscape because of its scale or character.

environments In reference to art work, the term refers to a construction or arrangement of things that establishes a special context for experiencing the creator's design. Environments may be extremely contrived or very simple, be made of fabrications devised by the artist or be merely a space he or she has defined as the environment of the work. Closely associated with *Conceptual Art* and *happenings*.

esthetic See *aesthetic*.

etching The process of producing a design on a metal plate by the use of acid or similar mordants. Also refers to a print made from an etched plate. To etch, then, is to expose a plate to such a mordant, known as "the etch."

expanded In typography, a modification of a typeface, the characters being wider than the standard version. The opposite of *condensed*.

Expressionism Strictly, German Expressionism. A *movement* in the arts that originated in Germany just prior to World War I, emphasizing the subjective aspects of the artist and subjects. By extension, any art of this kind.

extrabold In typography a very heavy version of a typeface, even bolder than *bold*.

eye level See Index.

fake duotone See *flat-tint halftone*.

fake process *Mechanical color separation* applied to reproduction in imitation of natural color. Rare today.

Fauvism A turn-of-the-century *movement* in France characterized by bright, flat zones of color. Similar to *Die Brücke* of Germany except that it took a joyous view of life.

feathering In *dry-brush* drawing, the use of a series of brief strokes to suggest the shaded edge of a form. In printing, the term refers to ragged edges in printed areas caused by poor ink distribution.

ferroconcrete Steel-reinforced concrete. That is, concrete reinforced by steel mesh, nets, rods, or bars that have been embedded in the cement while it was wet.

flat A grouping of film negatives, in register, attached to *goldenrod* or similar masking material. Placed against the plate and exposed to *arc* light, the flat conveys a positive image to the plate which can then be *etched* and prepared for printing.

flat-bed cylinder press   See Index.

flat-tint halftone   A black *halftone* printed on top of a flat tint of another color. Also called a "fake *duotone*."

flexography   Aniline printing. Used widely in the packaging industry, this process uses wraparound *plates* made of soft rubber or plastic and very fast-drying inks. It is well-suited to printing upon irregular surfaces.

flop   To turn a *negative* over so that the image is transposed.

Florentine Renaissance   The period from the fourteenth to the sixteenth century in the city of Florence, Italy, the cradle and jewel of the Italian *Renaissance.*

fluorescent ink   Inks containing substances that reflect ultraviolet light, producing an impression of extreme brilliance. Day-Glo is the best-known brand name of such an ink.

folio   Page number. Also, a large sheet of paper folded once in half. See Index.

folk art   Art and craft objects produced by untrained people as an expression of community life. Usually these artists are anonymous, the objects relatively utilitarian, and the community expression unconscious.

font   All of the *characters* of one size of a typeface.

foot   The bottom of a page.

form   In the fine arts, generally, the overall character of a work of art with regard to its appearance. In printing, the term refers to type and *engravings* locked into a *chase* for letterpress printing.

format   The overall appearance of a piece of printed work.

Formatt   Brand name of a cut-out *transfer type.*

foundry type   Metal type cast into hard metal and used in hand *typesetting.*

fountain   The ink reservoir of a mechanical printing press. It is from such fountains that ink is fed onto rollers and thence to the *plate.* In *offset lithography* the term is also used to designate a reservoir that holds a fountain solution (made of water or alcohol, acid, gum arabic, and a buffer) used to dampen the plates so that the blank areas will not accept ink.

French curves   A drafting tool made of a single piece of plastic about 1/16th inch thick, it contains a variety of curves to guide a pencil or pen. (The curves were not derived from intuitive notions about movements; they are expressions of mathematical formulae.) Many sizes and shapes are available.

fresco   The term *fresco,* used alone, refers to what is strictly *buon fresco*—a painting made on wet plaster so that the pigments become incorporated into the plaster. *Fresco secco* is painting done on dry plaster. Buon fresco is a very permanent form of *mural* painting; normally, the picture lasts as long as the wall stands.

frisket   Any thin film that can be used to mask out areas to protect them; it can later be removed. Used in *airbrush* work. Also the term in *letterpress* printing for paper used to cover a part of the *plate* so that it does not print.

frontality   In sculpture and figurative painting the term refers to the deemphasis of the lateral aspects of things. In painting, generally, it refers to arrangements of planes parallel to the picture surface. With respect to modern art it has a less precise meaning but suggests the emphasis of the tangible, two-dimensional surface of the picture.

frottage   See Index.

furniture   Pieces of wood, metal, or plastic having a rectangular configuration and a height just below that of type. They are used to fill in areas around type and *photoengravings* when these are *locked up* in the form.

Futurism   An Italian *movement* originating prior to World War I which hoped to glorify the dynamism of the machine age.

galley   A shallow tray used to hold type forms not in use.

galley proof   An impression of type to be checked for errors. Normally, a galley proof will not be spaced out precisely and is not fully assembled. Casually, called "galleys."

gang printing   Printing various jobs on the same sheet of paper. The paper is then cut up so that the jobs are separated.

genre   Normally, this word is used to refer to pictures in which the subject matter is drawn from everyday life. It is also used to describe whole categories of subject matter. Thus, one may speak of the "landscape genre," the "still life genre," the "genre of figure painting" and so on. Sometimes it is applied to media: genre of painting, sculpture, and so forth.

geodesic dome   A building form constructed of small, triangular *modules,* devised by Buckminster Fuller.

German Expressionism   See *Expressionism.*

gesso   A mixture of plaster of Paris and glue used as a painting *ground.*

glaze   In painting, a *pigment* mixed with oil and other fluids, applied in transparent layers. In *ceramics,* a vitreous coating applied before firing to seal the surface or used as a decoration.

Golden Age   Any period of high culture in a civilization. Most commonly, the term is applied to

Athens under Pericles (495–429 B.C.) and the Byzantine Empire under Justinian (A.D. 527–565).

goldenrod　A sheet of yellow-orange paper that is highly opaque and is used as a basis for *stripping* negatives.

golden section　A ratio between the two dimensions of a plane figure or the two divisions of a line such that the shorter element is to the larger as the larger is to the whole.

gouache　Opaque rather than transparent *watercolor*. The medium is essentially the same as in transparent watercolor, except that the proportion of binder to *pigment* is greater and an inert pigment such as precipitated chalk has been added to increase the opacity of the paint.

graphic　Literally, "written, drawn, or engraved." In the arts the term refers to drawing and printmaking. It is also used in the more commonplace sense to refer to something that is striking in its clarity.

gravure　A printing technique that is the commercial version of *intaglio* printing (such as line engraving or etching in fine art). In itaglio printing the areas that carry the ink to the paper are lower than the surface of the *plate*. More expensive and capable of subtler effects than *letterpress* or *lithography,* it requires better grades of paper for printing. The most familiar products printed in this manner are currencies.

gray scale　A set of graduations of values in 16 or 21 steps from white through grays to black. The steps are determined logarithmically rather than arithmetically and are used in processing photographic materials.

grippers　The mechanical clamps or "fingers" that hold paper onto the impression cylinder of a press during printing and release it when an impression has been run.

ground　Usually, the surface to which paints are applied. In *etching* the term refers to the waxy coating used to cover the plate and prevent mordants from acting upon it. Also, background, as in "figure and ground."

gutter　The blank space where two pages meet at the binding or centerfold of a book, magazine, or pamphlet. See Figure 312 for an illustration.

halftone　See Index.

halftone screen　See Index.

happening　A satiric act which takes place in a specially constructed or devised *environment.* Has both sculptural and theatrical aspects, but no permanent form is established. The "put-on" side of such an event is considered a vital part of its seriousness.

hard-edge painting　Any painting style employing very clean, sharp edges and flat areas of color. Normally restricted in application to work done by artists active after 1950.

head　The top of a page.

high key　Describing a picture or design in which the majority of *values* are lighter than middle gray.

highlight　In *chiaroscuro,* the reflection of the light source upon an object.

holding lines　These are lines drawn on the *mechanical* to indicate the precise zone that will be used for *halftone* or a *tint* or area of color.

hologram, holography　A hologram is an image produced by the simultaneous intersection of three laser beams. It is recorded on a piece of film and, when properly viewed, has all of the spatial qualities of the visible third dimension; overlappings change as the viewer moves relative to the transparent "window" of film exactly as they would if they were solid objects being examined. At one time, all holograms had to be seen under monochromatic light but there are now "natural light" holograms.

hue　The name of a color such as red, blue, yellow (primaries); orange, green, violet (secondaries); and the intermediate (tertiary) colors.

icon　Literally, in Greek, an image. Specifically, an image of a sacred person regarded as an object of veneration in the Eastern Church.

iconoclast　Image-breaker. Originally, the term was applied to the Byzantine ruler Leo III (680–740), who opposed the religious use of images. It has been extended to apply to anyone who actively questions generally accepted intellectual, ethical, or moral attitudes.

iconography　The study of images, primarily in terms of their symbolic intentions. The study of the deeper significance of content and its general importance is sometimes referred to as "iconology."

illuminated manuscript　A manuscript whose pages (and especially the initial letters) are decorated with silver, gold, and bright colors. Sometimes such manuscripts contain *miniatures,* but the illumination is in the lettering.

illustration　Imagery that relates specifically to something else. Illustrations serve to clarify or adorn an anecdote, literary work, description, event, and so on. The term is normally used to describe pictures on a comparatively small scale and most often is used in connection with pictures related to and appearing with a printed text.

impasto　The *texture* of paint applied in a thick, pasty form.

imposition　The way pages are set up in a press *form* so that after printing, *backing up,* folding, and trimming, the pages will be in the correct order.

impression　The printing of a single copy from a

plate. A specified set of such impressions is called an ''edition.''

Impressionism Specifically, French Impressionism. A *movement* in painting that originated during the last third of the nineteenth century. It attempted to attain a sort of ultimate *naturalism* by extending *Realism* beyond value relationships to an exact analysis of color. A typical Impressionist work is painted with short, brightly colored dabs of color, and use of *broken color* is frequent. Subject matter is unproblematical and undramatic.

indirect processing A form of four-color process in which the original *copy* is first *separated* into four *continuous-tone negatives.* These negatives are then screened separately. (In ordinary four-color processing the copy is screened and separated in one step.) The indirect method is slower but it has several advantages. Color can be corrected by *retouching,* for example. The major advantage is that the unscreened negatives can be enlarged or reduced to make any number of different reproductions without having to go back to the original copy.

intaglio A design sunk into the surface so that the impression it makes is *embossed,* in relief. Signet rings are common examples. In the fine arts *engraving* and *etching* are examples, in commercial printing *gravure.*

intensity The brightness or dullness of a color. Not to be confused with *value,* meaning the color's darkness or lightness. Often referred to as ''chroma.''

International Style A form of architecture dating from the third decade of the twentieth century. So-called because it hoped to rise above national peculiarities by being based upon science and reason. *De Stijl* and *Bauhaus* are associated with it. Stylistically, the designs tend to be geometric and unadorned, featuring a lot of glass and smooth wall surfaces. Walter Gropius, Le Corbusier, and Mies van der Rohe are important members of the *movement.* Also a term applied to a style of painting which arose at the end of the fourteenth century in France and Burgundy and is more accurately called ''International Gothic.'' It is a medieval-looking manner that entails the use of realism in details of landscape, animals, and costumes. Spreading rapidly throughout Europe, this style is represented in Burgundy by the Limbourg brothers, in Germany by Lochner, and in Italy by Gentile de Fabriano and Pisanello.

isometric projection A form of *axonometric projection.*

italic Slanted lettering, *like this.*

job press Often, ''jobber.'' A platen press (see Index) that is used to print small runs.

justify To make each line of type come out flush on the right, the left, or (as in this book) both, by adjusting the space between words and letters to make the lines even. A printed page with justified margins has lines of type that are even on both sides. Typewritten pages usually have *unjustified* right-hand margins.

keyline A *mechanical.* So called because of ''key lines,'' common to them. These are lines the designer has drawn around areas to show where a panel, color tint, or halftone is to be positioned. (See *holding lines.*)

kid finish Paper surface that is dull and smooth, having a slight roughness resembling the texture of kid leather.

kinetic art Works of art that are designed to move, either in response to human presence (because of touch, electric-eye beams, etc.) or because they are mechanized.

layout The plan of a design, showing the positioning of the various elements. Preliminary to a *comprehensive,* from which it is distinguished by tentativeness and lack of finish.

Letraset Brand name for a line of rub-off *transfer type.*

letterpress Printing method used to print from type or from *relief plates* or blocks. See Figure 278 for an illustration of the process.

light In typography, a modification of a standard type face so that the individual elements are leaner than in the original.

line conversion The modification of *continuous tone copy* into line copy by the use of special-effect *screens* that simulate *cross-hatching, stippling,* crayon drawing, fabric textures, line *engraving,* and other patterns.

line copy *Copy* without grays, drawn in black on white as *camera ready.*

line cut A *plate* that prints a reproduction of *line copy.*

lithography A printing method that takes advantage of the antipathy of oil and water to print from greasy images drawn upon limestone slabs. See Figure 280 for an illustration of the process. Also see *offset lithography.*

local color The natural color of an object as it appears to the normal eye at fairly close range under normal daylight. Contrast *optical color.*

lock-up In *letterpress,* the positioning and securing of type, *cuts,* and *furniture* in a *chase* for a *press run.*

logo or **logotype** Technically, this is a trademark or emblem made up of type characters linked to a common unit so that it can be printed as easily as a single piece of type. (Not the same as a ligature, which consists of two or more normally connected characters.) The term has come to mean, everywhere except where it is emphatically technical, a special design representing a company, organization, or publication.

lower case   Small letters as opposed to capitals.

low key   Opposite of *high key*. Describing a design or picture in which the majority of values are darker than middle gray.

magenta   Process red.

Magic Realism   A name for a style of painting that often resembles *Surrealism*. It involves a sharp, precisely detailed rendering of real-looking things. There are actually two varieties: (1) that which brings prosaic things into unusual juxtaposition, and (2) that which depicts ordinary things with such abnormal clarity as to lend them extraordinary overtones.

manuscript   In art, a hand-lettered book.

maquette   A scale model for a large sculpture. In Italian, *bozzetto*.

margin   The external border of a page.

mat   A decorative cardboard frame for display and support of a piece of art work.

mechanical   See Index.

medieval period   See *Middle Ages*.

medium   The vehicle or liquid with which *pigment* is mixed. In the more general sense, the material through which an artist finds expression (e.g., wood, paint, metal, fiber). In technical usage the term refers to the substance used to thin or otherwise modify the pigment and its vehicle.

Metaphysical painting   An Italian *movement* in painting during the earlier part of the twentieth century. The strangely evocative pictures by Giorgio de Chirico formed the basis of the movement; his paintings had a profound influence on *Surrealist* imagery.

Middle Ages   The period in Western history between about A.D. 400 and 1300.

military projection   A form of *planometric projection*.

mimesis   Literally, to imitate. Thus, the representation by illusion of the properties of the external world.

miniature   Any small image, of course, but particularly a little picture illustrating a *manuscript*, usually an *illuminated manuscript*.

minimal art   Painting and sculpture that stresses the simplest color relationships and/or the most clearly geometric forms.

modeling   The forming of three-dimensional surfaces. Thus, the creation of the illusion of such surfaces within the two-dimensional confines of painting or drawing. Also, of course, posing for a picture or sculpture.

module   A given magnitude or unit in the measurements of a work of art or a building. Thus, if one used a 16-inch module in designing a house, the dimensions would all be some multiple of 16 inches.

monochromatic   Consisting of variations of a single *hue*. (Individual color separations are monochromatic; when they are all combined, the result is polychromatic, that is, many-colored. See Colorplate 22.)

montage   See Index.

mosaic   The decoration of a surface with small pieces of stone, glass, or ceramic (called *tesserae*) set into cement.

movement   A general cultural tendency, sometimes carried out by people known to one another but frequently a more diverse response to given stimuli. Thus, one might speak of "the *Impressionist* movement," of a "Black Separatist movement," or of a "youth movement."

Multilith   Brand name of a little *offset* press used to print small jobs. It is manufactured by the Addressograph-Multigraph Company.

m weight   A measure that is exactly one half that of *basis weight*. It is the weight of 1,000 sheets of paper stock of a given size. Written as 50M, 75M, 78M, 100M, and so on.

Nabis   A group of French painters active between 1889 and 1899. The most important are Pierre Bonnard and Edouard Vuillard.

naive art   Works of art created by those without professional training in which a lack of sophistication is preserved as a positive value.

Naturalism   In literature the term corresponds to French *Realism* in painting. That is, it designates a *movement* having some of the same aims. In art, however, it is generally taken to signify accurate transcription of nature. William Harnett, Gustave Courbet, Edouard Manet, and Richard Estes are all naturalistic to some degree.

negative   A reverse image in which the value or hues are reversed. Also, a shortened version of "film negative," the name for the film reversal of *copy* (in which what was white becomes opaque and what was black becomes transparent) used in photomechanical processing.

negative areas   The shapes created in the background field when a figure is positioned against it. For example, a black dot on a white background not only produces the positive form of the solid black circle seen against white, it simultaneously produces the negative area of the white with a black zone taken out of it.

Neoclassicism   The attempt in the eighteenth and nineteenth centuries to revive the ideals of the Greeks and Romans of ancient times in French and German painting. By extension, any style based on similar principles. It attempts to attain perfection and harmony through prescribed limitations. Only certain subjects, modes of drawing and painting, color, and finishes are considered worthy of really serious painting and sculpture. French artist Jacques Louis David is the most famous of the Neoclassical painters. In architecture, the style depends upon deriving motifs from ancient examples. The Lincoln Memorial in Washington, D.C., designed by Henry Bacon, is a very late (1914–1921) example.

Neoplasticism   A type of *nonobjective* painting that reduced form to horizontal and vertical movements and used only black, white, primary colors, and (at the very end) neutral gray. Major practitioner was Piet Mondrian.

Neorealism   A contemporary *movement* featuring extremely realistic imagery, usually but not always derived from photographs. The common denominator of Neorealistic artists is a kind of cold detachment quite different from the mood of old-fashioned *Realism*. Some of it is similar to *Magic Realism* in its hardness, but the subject matter is matter-of-fact. Sculpture in this vein is apt to be polyester and fiberglass figures of an extreme, waxworks-like naturalism. A subcategory of Neorealism is *Photorealism*.

Neoromanticism   A painting style associated with and resembling the representational form of *Surrealism* but having more in common with *Metaphysical painting*. It is distinguished by the evocation of lyrical and nostalgic sentiment and by rather theatrical effects. The main practitioners of the style were Eugene Berman and Leonid Berman (who signs his work "Leonid"), Christian Bérard, and Pavel Tchelitchew.

Neue Sachlichkeit   "New Objectivity." A reaction against *Expressionism* that occurred in Germany during the Post-World War I period. It is "objective" only in that it took a more realistic view of the physical world. It was marked by a concern with social problems, and its most noted artists were Georg Grosz, Max Beckmann, Otto Dix, and Kathe Kollwitz.

New Objectivity   See *Neue Sachlichkeit*.

newsprint   An inexpensive grade of paper used for newspapers and other low-cost jobs. The *basis weight* is between 30 and 45 pounds. It has a coarse, absorbent surface.

nonobjective art   See *nonrepresentational art*.

nonrepresentational art   Works of art that make no attempt to produce illusions of reality.

Nordic Modern   The authors' name for the general trend in design represented by such things as the *Bauhaus, International Style* architecture, rationalized graphic design, the Helvetica typeface, and so on.

oblique projection   A form of engineering drawing in which one elevation is parallel to the picture plane while the plan and other elevations are set at an angle to it. See Figure 225 for an illustration.

offset lithography   The commercial form of lithographic printing. Very often referred to simply as "offset." See Figure 28 for an illustration of the process.

ogee   A double or S-shaped curve. Also refers to the bulging form produced by two such curves that mirror one another and meet in a point, as in an ogee arch.

op art   "Optical art." Works of art that depend for their interest upon optical illusion, fugitive sensations, and other subjective visual phenomena.

optical color   The apparent color of an object as opposed to its *local color*. For example, red looks black under a green light, pink looks white under a red one, and things very far away are turned blue by atmospheric refraction.

orthographic projection   The description of a three-dimensional object by a plan and elevations at right angles to one another and which, themselves, result from lines of projection that are perpendicular to the plane of the projection. See Figures 222 and 223 for illustrations.

overlay   A sheet of transparent or translucent film or paper placed over the base sheet of a *mechanical*. It may be for protection, may be part of the art work, or may carry instructions for the platemaker.

pagination   Numbering of pages in consecutive order.

palette   The surface on which a painter mixes paints. More generally, the habitual set of colors used by a given artist or group.

papier collés   Literally, "stuck paper" in French. Paper *collage*. Specifically, the collages Picasso and Braque did in Paris beginning in 1911.

pastel   *Pigments* mixed with gum and compressed into stick form for use as crayons. Also a work of art done with such pigments.

patina   A greenish film that forms on copper or bronze through oxidation. Effects similar to this on other substances.

penumbra   In *chiaroscuro*, the zones of transition from the lighted areas to the darkest (*umbra*).

permanence   In painting, the ability of a *pigment*, *vehicle*, or *medium* to resist change over a period of time.

perspective   The common name for *central projection*, a scheme for representing three-dimensional objects on a two-dimensional surface in terms of relative magnitude.

photoengraving   In commercial printing, a relief printing *plate* used by *letterpress*. Actually, "photoetching" would be a more accurate descriptive term, since the recesses are etched away with acid. (In the fine arts this is called "relief etching.") The term stems from the fact that photoengraving replaced *wood engravings* (themselves relief prints) for printing illustrations.

photomontage A montage of photographic images.

Photorealism A contemporary style in painting which derives its imagery from photographs of things and in which the photographic qualities are sedulously preserved, though usually enlarged enormously. Reproductions of the works of the Photorealists are usually indistinguishable from reproductions of photographs, and the works cannot be very accurately imagined from looking at pictures of them. This paradox is, of course, part of the appeal of the style.

photostat Brand name for a specific kind of photographic print, often referred to as a "stat." Original *copy* is photographed to make a paper negative from which a positive print is made.

photovoltaic Having the property of emitting electrons when stimulated by light. Photovoltaic cells are capable of generating electricity directly because of this phenomenon.

pica The most common typographic unit of measure. One pica is 1/6 inch wide. To each pica there are 12 *points,* and 6 picas make one inch. Also, typewriter type that is 10 characters to the inch as contrasted with the smaller *elite.*

pigment A substance that has been ground up into a fine powder and is used to color paints or dyes.

pitch A unit of width in typewriter type. It is based on the number of letters to the linear inch. *Elite* is 12 pitch, *pica* is 10 pitch.

planometric projection A form of engineering drawing in which the plan is parallel to the picture plane and the elevations at an angle to it. See Figure 226 for an illustration.

plate The metal or plastic surface that has been treated so that it will carry an image. More properly, a printing plate.

plate finish Paper surface that is hard and slick. Compare *kid finish.*

platemaker A technician who prepares *plates* for printing.

platen press See Index.

point Smallest typographic unit. It is 0.01383 inches, or about 1/72 of an inch.

Pointilism The application of pigment in small dots rather than brush strokes. Employed by Georges Seurat and his followers, but also used by other artists.

polyptych A many-paneled painting, sometimes hinged so that it can be folded up.

pop art Serious art work using, as subject matter, elements from commercial illustration, cartoons, signs, and ordinary mass-produced objects.

Postimpressionism A catchall term for the styles employed by painters influenced by *Impressionism* who modified it into more personal modes of expression. The painters most frequently cited as examples of Postimpressionists are Cézanne, Renoir, van Gogh, Gauguin, and Seurat. The term itself was invented by British art critic Roger Fry for an exhibition in London in 1910–1911.

post-modernism A term used to describe a contemporary reaction against *Nordic Modern* design. It is not anti-modernist since it makes no attempt to hold up tradition as the objective standard of the style, yet it is quite different from standard modernism in being willing to use things for mere effect.

pottery Objects of clay that have been hardened by firing.

Precisionism A *movement* in American art that began in the 1920s, typified by the work of Charles Sheeler and Georgia O'Keeffe. It combines realistic subject matter with extreme formal control.

press run The number of sheets printed.

Prestype Brand name for a line of rub-off *transfer type.*

primitive art A term becoming less common because of its pejorative overtones and racist connotations. It has the following meanings, more or less in this order: (1) art produced by tribal cultures; (2) art created by amateur artists whose unsophisticated vision or lack of technical skill have, for one reason or another, come to be counted as virtues; (3) art produced by painters active before 1500 in the Netherlands and Italy.

print A work of art produced in multiple copies by hand methods and printed by the artist or under the direct supervision of the artist.

process camera See Index.

program The collection of instructions and the operational routine necessary to direct an intellectual activity mechanically, particularly by a computer.

progressive proofs Sample runs pulled of color plates showing each color in isolation and in combination with the others. Colorplate 22 is such a run.

projection See Index.

proportional scale An instrument designed to show the measurements of any rectangular shape in terms of percentages and inches as it is enlarged or reduced in size. See Index.

pulp paper A low-grade paper that is coarser and bulkier than *newsprint.* Used for very low-cost printing. During the 1920s, 1930s, and 1940s, this rough *stock* was used for printing science fiction, detective, and adventure magazines and other cheap, popular literature. Consequently, such magazines are known as "pulps."

push pin   A tack used by graphic artists to hold work to drawing boards and tackboards. It has a cylindrical head that makes it easy to use without damaging surfaces around it.

Realism   A nineteenth-century *movement* in painting which stressed matter-of-fact descriptions of actual things. Frequently, as in the work of Courbet (who coined the name of the style), Realism focused on the squalid and depressing. But Manet's work, and much of Courbet's for that matter, shows that ignoble themes are not fundamental to the manner. It is really a nonsentimental version of *Naturalism*.

ream   500 sheets of paper.

recto   The right-hand page of a book lying open. Opposite of *verso*.

Reformation   A religious revolution in western Europe during the sixteenth century. It began as a reform movement within the Roman Catholic Church but evolved into the doctrines of Protestantism. Its outstanding representatives were Martin Luther and John Calvin.

refraction   The bending of a light ray or wave of energy as it passes through a substance.

Regionalism   The name given to the work of certain American painters prominent in the 1930s. Their work tended to represent specific regions and to portray common people in everyday activities. Notable among these are Grant Wood (Iowa), Thomas Hart Benton (Missouri), John Steuart Curry (Kansas), Edward Hopper (New England), and Peter Hurd (Southwest).

register   Positioning of one thing over another so that the elements of each have a correct relationship.

registration marks   Cross hairs used to position overlays and negatives in *register*.

relief printing   Describing any form of printing that uses a raised surface to carry ink to the material that is printed upon. *Woodcut* in the fine arts and *letterpress* in commercial printing are the most common kinds of relief printing.

Renaissance   Usually signifies the time of the rebirth of art and humane letters (as opposed to divine letters) during the fourteenth and fifteenth centuries, particularly in Italy. Early Renaissance covers the period from about 1400 to 1500 and the High Renaissance from about 1500 to 1527 (when Rome was sacked by the Holy Roman Emperor Charles v). These dates are for painting and architecture in Italy; in literature the beginning date is often given as Petrarch's birth in 1304. Some authorities define the High Renaissance as lasting until 1580. With respect to England, the dates are quite different; William Shakespeare, who was born in 1564 and died in 1616, is considered a Renaissance playwright. In other words, it's confusing.

repoussé   The forming of a design in metal by working it from the back and leaving the impression on the face. Compare *chase*.

reproduction   The production of multiple images of a work of art by photomechanical processes. Compare *print*.

retouching   Correcting or otherwise altering a photograph before it is reproduced.

reverse   Technical jargon for a printed image in which the blacks and whites are exactly the reverse of the original *copy*. In other words a printed negative. To achieve a reverse, the negative of the shot is used to make a positive on film instead of paper and this is then used to produce the *plate*.

rococo   Primarily a style of interior decoration in vogue in France from the death of Louis xiv in 1715 to about 1745. It had some influence in Italy and Spain but was most influential in northern countries. Tending toward prettiness and frivolity, it influenced such painters as Watteau, Boucher, Fragonard, and Tiepolo, but it reached its loftiest form in Catholic Germany and Austria, where it produced extraordinarily beautiful church interiors.

Romanticism   A *movement* in painting that first appeared in France during the early nineteenth century. It attempted to express the entirety of human experience, both real and imagined. By extension, the term is applied to any work of art expressing interests thought to be similar to the characteristics of Romanticism. Anticipated in literary works by Byron, Goethe, Wordsworth, Coleridge, and others, Romantic painting frequently drew its subject matter from the works of these authors. In music, however, Romanticism is a later phenomenon, identified with Schumann, Chopin, Wagner, Berlioz, Liszt, and Mendelssohn, among others.

rotary press   See Index.

rough   A sketch that gives the general notion of the design. Small, tentative roughs are called *thumbnail sketches*.

R-value   Resistance to thermal conductivity, a coefficient number used to indicate the effectiveness of insulating material with respect to blocking the flow of heat. The higher the number, the more effective the insulation. The highest conventional insulating material is at R-38 for 12-inch material. A rating of R-14 is becoming fairly standard in new homes today.

saddle-wire stitching   The common way of binding magazines and pamphlets by driving wire staples through the *backbone* into the center-spread, where they are clinched. Compare *side-wire stitching*.

sandcasting   See Index.

sans serif   Lettering without *serifs*.

scaling   Calculation of the amount of enlargement or reduction art work is to undergo for reproduction. Usually expressed in form of percentages. (The original art work is thought of as 100 percent. A 50 percent reduction of a design drawn 8″ × 8″ would be 4″ × 4″. A 75 percent reduction would be from 8 to 6 inches. A 25 percent reduction, then, would take the sides down to 2 from the original 8 inches. If the design were enlarged from 8 to 10 inches, the reproduction would be at 125 percent.)

school   A word with a great variety of meanings, all of them based on the assumption of identifiable similarities among the works of various artists who (1) studied with the same master, (2) imitate the same master, (3) work together as a group, (4) express the same interests in their works, (5) lived in the same place and time, or (6) have the same country of origin. Thus: the School of Raphael (1 and 2); the French Impressionist School (3); the Pop Art School (4); the School of Florence (5); the Italian School as contrasted with the French School (6).

screen   Usually, *halftone screen*. Sometimes *silk screen*. (These meanings apart from such obvious ones as the use of wire screening in ceramics and sculpture to filter out foreign substances and large particles.)

screen process printing   See *silk screen*.

sculpture in the round   Freestanding statues as opposed to relief sculpture.

separation   See Index.

serif   See Fig. 291.

serigraph   A fine arts print done by the *silk-screen* printing method and printed by hand.

sgraffito   A design produced by scratching through one layer of material into another one of a contrasting color. Frequently used in pottery decoration but also employed by painters and sculptors.

shade   In color theory, to darken a hue by adding black to it.

shading sheet   See Index.

side-wire stitching   Method of binding books or pamphlets by driving wire staples through the binding edge from the top sheet to the back sheet where they are clinched. Compare *saddle-wire stitching*.

silhouette halftone   A *halftone* in which the main image has been set off so as to be isolated from any background or framing rectangle by elimination of the dots surrounding it. Sometimes called an *outline halftone*. See Figs. 362, 363, and 364 for examples.

silk screen   Often today called "screen process printing." A printing process in which ink is squeezed through a screen of silk that has been tightly stretched over a frame and has been masked in places by a stencil. Photographic stencils may be used, even in *halftone* form, as long as the dot pattern is larger and coarser than the mesh of the silk. Silk screen permits an extremely heavy application of ink or paint on practically any material. Posters are the graphic designs most often produced by this process, but it has become popular as a means of imposing designs on fabric, particularly tee-shirts. Printing can be done manually or by mechanical process.

simultaneous contrast   See Index.

slip   Clay that has been thinned with water to the consistency of cream. It can be poured into molds of plaster and thereby cast into pots or can be used as a paint to ornament *ceramic* pieces.

Social Realism   A term used to describe various styles in art which emphasize the contemporary scene, usually from a left-wing point of view and always with a strong emphasis on the pressures of society on human beings.

solar gain   The increase of temperature caused by the rays of the sun falling upon something.

split-fountain printing   A printing technique that involves using one color in the left-hand side of the ink *fountain* of a press and another color on the right-hand side so that the left prints a different hue than the right. As the rollers turn, they blend the colors into a continuous gradation, producing a third color wherever the two original hues fuse. By using several dividers to separate pools of ink in the fountain, more than two colors can be used to produce rainbow, spectrum effects.

square halftone   Any rectangular *halftone*. So called because the edges of the *plate* are squared up on a special machine.

staining power   The capacity of a pigment to affect other pigments with which it is mixed in terms of hue character.

stereotype   In common parlance, a standardized mental picture that represents things in a false and oversimplified manner. (For example, the stereotype of women as poor drivers, of black people as lazy and superstitious, or of professors as . . . as—something or other. . . .) In art the term refers to a form that is repeated mechanically, without real thought. In printing it refers to identical relief printing surfaces produced by taking a mold from a master surface and then casting copies. These copies are called stereotypes.

Stijl, De   A Dutch magazine of the early twentieth century devoted to *Neoplasticism*. In Dutch the title means "the Style." The term is used to refer to the ideas advocated by the magazine; they had a marked influence on the *Bauhaus* and upon graphic design and architecture generally.

stippling To draw by making dots instead of lines.

stock Any printing paper.

stripping Securing the assembled negatives to the *flat*. The person who does this is called a stripper or makeup person.

stylize See Index.

subtractive color See Index.

Suprematism A *nonrepresentational* style devised by Kasimir Malevich in the early twentieth century. It is a sort of Russian *Neoplasticism*.

surprinting Combining images from two different negatives by *double-burn*, especially when a line image, such as lettering, seems to have been superimposed upon a *halftone*. Compare *drop out*.

Surrealism The post-World War I *movement* which drew inspiration from Freudian psychology and extended the arbitrary irrationality of *Dadaism* into a doctrinaire exploration of the unconscious.

swipe The appropriation of an existing graphic image for use in one's art work. While this may seem to smack of theft, the practice is general throughout the field of graphic design and is accepted. Thus, it is common to derive backgrounds from photographs, and illustrators lift one another's poses and lighting effects. What must be avoided at all costs is plagiarism, the actual copying of another's work. To derive an original image from another one is permissible; to imitate an original image is illegal.

synesthesia The psychological process whereby one kind of stimulus produces a secondary sensation of a different sort, as when a color suggests an odor or a sound a color.

systemic Relying upon a system or set of rules.

tapestry A heavy, handmade textile in which the threads (usually the weft or horizontal threads) are woven to create a design.

tempera Paint using egg as the binder for the pigment. Technically "egg tempera." It is highly *permanent*.

texture The tangible quality of a surface, that is, its smoothness, roughness, slickness, and so on. Simulation of same in drawing or painting. See Index.

thermography A process used to imitate the *embossed* effect of true hand steel engraving by producing raised letters on a sheet of paper. It entails dusting the printed sheet, while the ink is still wet, with a resin powder that adheres to the ink. When the sheet is passed through a heating unit designed for the purpose, the particles of resin fuse with the ink and cause it to swell up.

thirty Newspaper people use the symbol "-30-" at the end of typed *copy* to indicate the end of the story.

thumbnail or **thumbnail sketch** A small, tentative *rough*.

tint To lighten a hue by introducing white into it. Also, in printing, a flat *halftone* of uniform character.

tondo A circular painting.

tone Synonymous with *value*.

transfer type See Index.

trimetric projection A form of *axonometric projection*.

triptych A three-panel picture, usually so designed that the central panel is twice as wide as the other two, with the latter hinged so that they can be folded over the central one.

trompe l'oeil Literally, "trick the eye." Extremely illusionistic paintings intended to lead the viewer into believing that all or a part of the image is not painting but actuality.

type-high The height of a standard piece of type. In the United States that is .918 inches.

type metal The alloy used to cast type. Basically, it is made of lead, tin, and antimony, and sometimes a tiny bit of copper.

typesetting Setting type up for printing. This can be done by hand, by machine, or by phototypesetting. A person who sets type by hand or by machine is called a "compositor," and typesetting is also referred to as "composing type."

typewriter composition Often called "direct impression" or "strike-on," this is the composition of *line copy* for reproduction by use of a typewriter.

umbra The darkest area of shadow in a *chiaroscuro* rendering. Also called the "core of shadow."

unjustified Describing lines of type aligned on the left or the right side but ragged on the other due to uneven length. Unjustified right-hand margins are characteristic of typewriting. Compare *justify*.

upper case The capital letters in a typeface.

value The relative darkness or lightness of a color.

vanishing point The point to which receding lines converge in *central projection*.

vanitas See Index.

Varityper Brand name of a unique typewriter that composes type in several styles and can *justify* right-hand margins semiautomatically. Manufactured by the Addressograph-Multigraph Company.

Velox A photographic print made from a *halftone negative*. Since it is *line copy* it can be used in a *mechanical* along with other line copy. The name *Velox* is derived from Eastman Kodak's trade name for glossy photographic papers.

verso The left-hand page of a book spread open. Opposite of *recto*.

vignette halftone A *halftone* in which the background or perimeter fades imperceptibly into the surrounding white of the paper. In other words, a half-tone with hazy edges. Done by *airbrushing*.

warm color   Red and the hues that approach red, orange, and yellow. Sometimes yellow-green is considered a warm color.

wash drawing   Watercolor-like rendering done with ink, a brush, and water to produce *continuous tone copy*. Compare *dry-brush*.

watercolor   Strictly, any pigment mixed with water, but usually signifying transparent watercolor in which the binder is gum arabic.

watermark   A faint, translucent design used as a trademark by paper manufacturers. It is produced by a relief pattern of wire soldered onto a wire-mesh roller (called a dandy roller) that is used in the initial forming of pulp into paper.

web   In printing, a continuous roll of printing paper.

web-fed   Descriptive of presses that print from continuous rolls of paper.

weight   Applied to type, this term has reference to the variations in thickness of the elements of a given typeface from *light* through regular to *bold*. In connection with paper measurements it is synonymous with *basis weight*.

woodcut   A relief surface for printing carved from the plank grain of a piece of wood. Also the image printed from such a surface. Also called "woodblock." See Figure 277.

wood engraving   A relief printing surface carved from the end grain of a piece of wood. The image printed from such a surface. The end grain has no lateral direction and permits much more detailed carving than plank grain; typically wood engravings are more detailed than *woodcuts*.

woodtype   Type made from wood. Normally used only for display lettering over one inch tall.

Zipatone   Brand name for a set of screen patterns printed on transparent film with an adherent backing. Used to add dark and light *values* and decorative detail to *line copy*.

bibliography

art and design in general

BANHAM, REYNER. *Theory and Design in the First Machine Age.* London: The Architectural Press, 1960.

BEVLIN, MARJORIE ELLIOTT. *Design Through Discovery.* 3rd ed. New York: Holt, Rinehart & Winston, 1977.

FAULKNER, RAY, and EDWIN ZIEGFELD. *Art Today: An Introduction to the Visual Arts.* 5th ed. rev. New York: Holt, Rinehart & Winston, 1969.

FELDMAN, EDMUND BURKE. *Varieties of Visual Experience: Art as Image and Idea.* Englewood Cliffs, N.J.: Prentice-Hall. New York: Harry N. Abrams, 1971.

GOMBRICH, ERNST H. *Art and Illusion.* New York: Pantheon, 1960.

ITTEN, JOHANNES. *Design and Form.* 2nd rev. ed. New York: Van Nostrand Reinhold, 1977.

KEPES, GYORGY. *Language of Vision.* Chicago: Paul Theobald, 1969.

MOHOLY-NAGY, LASZLO. *Vision in Motion.* Chicago: Paul Theobald, 1947.

RICHARDSON, JOHN ADKINS. *Art: The Way It Is.* 2nd ed. rev. Englewood Cliffs, N.J.: Prentice-Hall. New York: Harry N. Abrams, 1980.

history of art and architecture

GOMBRICH, ERNST H. *The Story of Art.* 12th ed. rev. and enl. London: Phaidon, 1972.

HARTT, FREDERICK. *Art: A History of Painting, Sculpture, and Architecture.* 2 vols. Englewood Cliffs, N.J.: Prentice-Hall. New York: Harry N. Abrams, 1976.

JANSON, H. W. *The History of Art: A Survey of the Major Visual Arts from the Dawn of History to the Present Day.* Rev. and enl. ed. Englewood Cliffs, N.J.: Prentice-Hall. New York: Harry N. Abrams, 1969.

PEVSNER, NIKOLAUS. *An Outline of European Architecture.* 6th (Jubilee) ed. rev. Baltimore: Penguin Books, 1960.

graphic design and illustration

BIEGELEISEN, J. I. *Screen Printing.* New York: Watson-Guptill, 1971.

CRAIG, JAMES. *Production for the Graphic Designer.* New York: Watson-Guptill, 1974.

GARLAND, KEN. *Graphics Handbook.* Paper. New York: Van Nostrand Reinhold, 1966.

LATIMER, HENRY C. *Preparing Art and Camera Copy for Printing.* New York: McGraw-Hill, 1977.

MANTE, HARALD. *Photo Design: Picture Composition for Black and White Photography.* New York: Van Nostrand Reinhold, 1971.

RICHARDSON, JOHN ADKINS. *The Complete Book of Cartooning.* Englewood Cliffs, N.J.: Prentice-Hall, 1977.

SCHLEMMER, RICHARD. *Handbook of Advertising Art*

Production. Englewood Cliffs, N.J.: Prentice-Hall, 1974.

Speedball Textbook. New York: Hunt Manufacturing Co., 1972.

Stone, Bernard, and Arthur Eckstein. *Preparing Art for Printing.* New York: Van Nostrand Reinhold, 1965.

The Type Specimen Book. New York: Van Nostrand Reinhold, 1974.

elements of art and design _____

Albers, Josef. *Interaction of Color,* rev. ed. New Haven, Conn.: Yale University Press, 1972.

Birren, Faber. *Color, Form, and Space.* New York: Reinhold, 1961.

Doblin, Jay. *Perspective: A New System for Designers.* 11th ed. New York: Whitney Library of Design, 1976.

Itten, Johannes. *The Art of Color.* New York: Van Nostrand Reinhold, 1974.

Mills, John. *The Technique of Sculpture.* New York: Reinhold, 1965.

Proctor, Richard M. *The Principles of Pattern: For Craftsmen and Designers.* New York: Van Nostrand Reinhold, 1969.

Selz, Jean. *Modern Sculpture, Origins and Evolution.* Trans. Annette Michelson. New York: Braziller, 1963.

Walters, Nigel V., and John Bromham. *Principles of Perspective.* New York: Watson-Guptill, 1974.

Watson, Ernest W. *How to Use Creative Perspective.* New York: Van Nostrand Reinhold, 1957.

index

Page numbers are in roman type. Figure numbers of black-and-white illustrations are in *italics*. Color-plates are specifically so designated. Names of artists whose works are discussed are in CAPITALS. Titles of works are in *italics*. An "n" or a "c" following a page or figure number indicates a note or caption respectively.

a

Abraham's Oak (Tanner), 118, *167*
Abstract Painting, Blue (Reinhardt), 118, 133, *168, 188*
Abstraction, 56, *57c.*
Academicism, 65, 81n.
Additive color, 120, 194, Colorplate 15B
ADES, CONSTANCE, *Sir Roger de Trumpington, 61*
Adoration of the Shepherds (Correggio), 148, *214*
Adrian van Rijn, 119c.
Advertising, 12, 65
Aerial photography, 3, 168, *2, 252*
l'Aficionado (Picasso), *126*
Afro-Americans, 39, 240
Alarm system, pressure sensitive, 245–46
ALBERS, JOSEPH, *Homage to the Square, 36, 54*
Alberti, Leonbattista, 177n.
Alhambra, 241, *372B*
All Combinations of Arcs . . . (Lewitt), 146, *208*
Alloway, Lawrence, 15
Amateur artists, 129
American Gothic (Wood), 59, *89*
Analogous colors, 116
Anamorphic distortion, 22, 29, 196, 239, *30B, 31, 33, 47*
Anamorphic Portrait of Edward VI (Scrots), 22, *31*
ANDERSON, DAN, Bottle, 138, *198*
Anecdotal painting, 81n.
Ansonia (Estes) 147, *209*
ANUSZKIEWICZ, RICHARD, 121, *Splendor of Red,* 121, Colorplate 17

Apartment housing, 223
Applied art, 3, 5
Architectural rendering, 160, *237*
ARRESSICCO, Arressicondo, 223, *344,* logotype, *10C,* program covers, *316A, 319,* record jacket *12,* underground house, 230, *351*
Arressicondo (Arressicoo), 223, *344*
Art and Camouflage (Behrens) 41, *64*
Art Decco, 10
Art Nouveau, 10
Artificial lighting, 98, 142, *140B*
AT&T Office Building (Johnson and Burgee), 218, 221n., *339*
Atmospheric perspective, 165, 176, *246*
At the Milliner's (Degas), 51, *92H, 93H*
Automobiles, 238n.
Autonomous Dwelling Vehicle (Bakewell and Jantzen), 230–32, 234, *352, 353*
Avant garde, 243
Axonometric projection, 154–56, *227, 228*
Ayer, A.J., 250n.

b

BAILEY, WILLIAM, 138, *Large Umbrian Still Life,* 138, *199*
BAKEWELL, TED, 230–32, 234, Autonomous Dwelling Vehicle, 230–32, 234, *352, 353*
BARNES, ERNEST, *Untitled, 151F*
Baroque, 53
Basket of Apples (Cézanne), 166-67, *247*
Bauhaus, 6, 10, 12, 236, 250n.

BEARDEN, ROMARE, 39, 41, 239–40, 242, *Prevalence of Ritual: Baptism,* 39, 41, 239–40, 242, *59, 368*
Bedroom at Arles (van Gogh), 3, 129, *6, 178*
BEHRENS, ROY, 41, 43, 129, illustrations, 41, 43, *63, 64, Art and Camouflage,* 41, *64*
BELL, ROBERT, illustration, *104B*
Benday patterns, 189
Bernard, Emile, 166
BERNINI, GIAN LORENZO, 95, David, 95, *137*
Bird in Space (Brancusi), 93, *135*
Black letter, 215n.
Black Yellow Red (Youngerman), 45, *49*
Booster (Rauschenberg), *106*
Borghese Palace, Rome, 95
Bottle (Anderson), 138, *198*
BOULDS, DAVID, 228–29, solar home, 228–29, *349*
Bozzetto, 95
BRANCUSI, CONSTANTINE, 93, *Bird in Space,* 93, *135*
Braque, Georges, 39, 167
BREUER, MARCEL, side chair, *8C*
BROD, STAN, exhibition catalogue, *152H*
BROWN, SCOTT, Lieb House, 218–21, *340, 341*
BRUEGEL, PIETER, 165, *Return of the Hunters,* 165, *245*
Brunelleschi, Filippo, 162, 177n.
Brunswick folding chairs, 232n.

Prentice-Hall, Inc., Englewood Cliffs, New Jersey 07632

ISBN 0-13-060186